RETHINKING THE WORLD

DISCOVERY AND SCIENCE IN THE RENAISSANCE

This catalogue was published with funds from the Wendell L. Willkie Educational Trust, administered by the Indiana University Foundation.

Cover illustration: No. 33. Cieza de León. *The silver mines at Potosí (in modern Bolivia).*

RETHINKING THE WORLD

DISCOVERY AND SCIENCE IN THE RENAISSANCE

An exhibition prepared and described by Helen Nader

The Lilly Library
Indiana University
Bloomington, Indiana
1992

ISBN: 1-879598-12-4

CONTENTS

FOREWORD

The books and manuscripts that have made this exhibition commemorating the five hundredth anniversary of Christopher Columbus's first voyage to the Americas possible are drawn from the collections formed by J. K. Lilly, Jr., Bernardo Mendel, and Charles R. Boxer. These three distinguished collections combine to provide a rich resource for the study of the era of European voyages of discovery and the colonies established by the Spanish, the Dutch, and the Portuguese.

Josiah Kirby Lilly, Jr., 1893–1966, grandson of the founder of Eli Lilly and Company, assembled a collection of English, American, and Continental literature, science, and Americana which came to Indiana University in 1956 to form the nucleus of the Lilly Library.

Bernardo Mendel, 1895–1967, businessman of Vienna, Bogotá, and New York City, collected books and manuscripts relating to the age of discovery, exploration, and settlement in the Americas with special emphasis on the Spanish empire. His collection is one of the most extensive in this area ever assembled by a single individual.

Charles R. Boxer, b. 1904, is both an internationally known historian and a highly knowledgeable collector with interest in the Dutch and Portuguese colonies. His collection took us into Africa, India, and the Far East.

It is because of the creative intelligence of these three collectors that the Lilly Library is able to provide such rich materials for the study of this period in the world's history. We salute them with this exhibition and its catalogue.

William R. Cagle
Lilly Librarian

Introduction

Five hundred years ago, in October of 1492, Christopher Columbus led a Spanish fleet to America and thus began the mutual discovery of peoples and cultures between the two hemispheres. This exhibition marks the quincentennial of the first Columbian voyage, which left Spain on August 3, 1492, and returned to Spain in March of 1493.

The 1492 voyage was not the first; others had crossed the Atlantic before. It was unique, however, because of its lasting consequences. The Columbus voyage initiated permanent contact between the Eastern and Western hemispheres. Except for 1497, ships have traveled between Spain and the Americas every year since 1492. Europeans completed dozens of transatlantic voyages before Columbus's death in 1506. In 1508 alone Spaniards made forty-five transatlantic voyages. On April 7, 1521, they finally achieved the original objective of Columbus's voyage: three Spanish ships commanded by the Portuguese Ferdinand Magellan reached Asia by sailing west. For the first time, the whole world was knit together.

Europeans eagerly absorbed new information about continents, islands, and societies they had never dreamed existed. For this information, they depended on reports written by explorers, colonists, and scientists in the Americas. These first-hand reports circulated throughout Europe in many languages, in the form of books and pamphlets, and stimulated new ideas. Hundreds of books published between 1493 and 1600 proposed solutions to the conflicts and questions raised by the European discovery of the Americas.

The Lilly Library is uniquely endowed with the materials to display these proposed solutions. Its collections contain many of the books about the voyages and the Americas published in the fifteenth and sixteenth centuries, including Christopher Columbus's first report on the Americas in 1493 and the earliest books printed in sixteenth-century Mexico and Peru. In these books, we can trace the establishment of modern America's diverse societies and understand the major issues that Europeans sought to resolve.

For many Europeans, expanding knowledge of the physical world and growing acquaintance with previously unknown native societies raised serious practical and theoretical issues. This exhibition traces the two major consequences—the emergence of America as a unique society and the transformation of European concepts of the natural world—as well as commemorating the Columbian voyages.

Helen Nader
Department of History
Indiana University

1.

Cristoforo Colombo, 1451–1506. *Epistola Christofori Colom . . .* Rome: Stephan Plannck, 1493.

During Columbus's return voyage in early 1493, violent storms blew *Niña* and *Pinta* off course. *Pinta*'s captain, Martín Alonso Pinzón, made harbor in the northern Spanish port of Bayona, while Columbus took refuge in Lisbon's port, Belém. Both captains immediately sent word to the Spanish monarchs of their return, and both put into the ships' home port of Palos on March 15, just a few hours apart. King Fernando and Queen Isabel immediately wrote to Columbus, inviting him to the royal court, which was in Barcelona at the time. In Barcelona, Columbus received a joyous welcome and turned his shipboard diary over to the monarchs.

Fernando and Isabel were determined that the Kingdom of Castile would have legal possession of the islands that Columbus had discovered. To achieve this, publicity was essential. First, they required a thorough and succinct description. For this, they turned to members of Fernando's staff, experienced in the drafting of legal documents, who culled passages from Columbus's letter and diary to write a description of the Caribbean islands. The monarchs had this first description of America printed in Barcelona. It has since been known as the *First Letter from America.*

The monarchs feared Portuguese incursions into the territory that Columbus had explored and claimed for Castile. Castile's maritime rivalry with Portugal had dragged on for nearly a century, and their ambassador in Lisbon reported alarming news: the Portuguese king was outfitting a fleet, destination secret.

By publishing Columbus's description written in the Castilian language, which was the *lingua franca* of international commerce in Portugal, Fernando and Isabel made sure that the Portuguese could not claim ignorance of where Columbus had traveled and what he had claimed for Castile. With the *First Letter from America,* in short, Fernando and Isabel gave notice that Columbus's discoveries were off limits to other nations, especially Portugal.

Fernando and Isabel next took steps to make sure that the whole European world knew and recognized Castilian sovereignty over the islands. They immediately sent the *First Letter from America* to their ambassador in Rome, Bishop Bernardino Carvajal. Carvajal had the letter translated into Latin by an Aragonese cleric in Rome, Leandro de Cosco. In Latin, the international language of Europe, the letter was printed by Stephanus Plannck.

2.

Cristoforo Colombo, 1451–1506. *Epistola Christofori Colom: cui etas nostra multum debet: de insulis Indie supra Gangem nuper inventis.* Rome: Stephan Plannck, 1493.
Fernando and Isabel made sure that Columbus's *First Letter from America* would be widely distributed and read throughout Europe. Probably at their initiative,

the book went through nine editions within a year: three editions in Rome, one
in Antwerp, one in Basel, three in Paris, and another in Basel in 1494. An Italian
translation in poetry by Giuliano Dati had three editions in 1493 (one in Rome
and two in Florence). A German edition appeared in Strassburg in 1497. A
second edition in Castilian appeared in Valladolid in 1497.

This copy of the second Plannck edition in Rome is open to the final page, which
describes Columbus as "Christopher Columbus, Commander of the Ocean
Fleet."

3.
Bernardino Carvajal, 1456–1523. *Oratio super praestanda solenni obedientia
sanctissimo . . . Alexandro papae VI . . .* Rome: Stephan Plannck, 1493?

In a further effort to assure Castilian sovereignty over the newly discovered
islands, Fernando and Isabel appealed to the fifteenth-century's equivalent of an
international arbiter, Pope Alexander VI. Their ambassador to the papacy,
Bernardino Carvajal, made this speech to the pope on June 19, 1493, in which he
presented Castile's claims and requested that Alexander confirm Castilian
possession of the islands.

The appeal was successful. Alexander issued a papal decree, *Inter caetera*,
confirming Castilian possession of the islands that Columbus had discovered
and any other islands or continent that Castilians might discover in the future
beyond an imaginary line in the Atlantic Ocean. This Line of Demarcation,
according to *Inter caetera*, was drawn from the Arctic or North Pole to the
Antarctic or South Pole, one hundred leagues south and west of the Azores and
Cape Verde Islands.

Spanish appeal to the papacy had the intended effect. Portugal, worried that the
papal decrees gave Castile a right to the eastward route to the Indies and to
lands off the West African coast, agreed to negotiate directly with the Spanish
monarchs. On June 7, 1497, Castile and Portugal signed the Treaty of
Tordesillas, setting the Line of Demarcation 370 leagues west of the Cape Verde
Islands. Through this compromise, Castile secured possession of Columbus's
discoveries, including eventually North and South America, while Portugal
retained rights to lands discovered east of this line, which included sub-Saharan
Africa and the East Indies, to which, one month later, the Portuguese king would
despatch Vasco da Gama.

4.
Carlo Verardi, 1440–1500. *In laudem serenissimi Ferdinand hispaniane regis Bethicae
& regni Granatae obsidio victoria & Triu[m]phus. Et de insulis in mari Indico nuper
inuentis.* Basel: Johann Bergmann de Olpe, 1494.

Many Spaniards had seen Columbus and his half-dozen Indian captives as they
traveled overland to Barcelona in the spring of 1492. Bartolomé de Las Casas
was a little boy in Seville at the time and years later remembered his excitement
at seeing the rubber balls carried by the Indians—the first rubber seen by

Europeans. He did not know then that these large rubber balls were used in a deadly form of soccer, in which the defeated player lost his life as well as the game.

In Barcelona, the monarchs authorized a second Columbus voyage to secure Castilian possession of the island of La Española (modern Haiti and Dominican Republic) and to resupply the thirty-nine men the admiral had been forced to leave behind after *Santa María* ran aground on Christmas day, 1492. They appointed the new deacon of Seville cathedral, Juan Rodríguez de Fonseca, to organize the voyage.

The outfitting of this fleet—which eventually grew to seventeen ships and about 1,200 men—progressed at a feverish pace during the summer of 1493. News of the Americas spread throughout southern Spain, where a large proportion of the population made its living by fishing and shipping or by producing supplies and equipment for seafaring voyages.

While thousands of Spaniards saw native Americans and their artifacts firsthand, northern Europeans had to depend on illustrated books. This situation led to many misperceptions.

Printers in northern Europe kept woodcut blocks in stock to illustrate their books. Most of the images had been created for earlier editions of romances, poetry, and moral tales; the artists had never seen the Americas, but the printers just kept reusing the old blocks or, when these wore out, commissioned new blocks with the same images. These images, which pretended to represent the reality of America but in fact were purely imaginative, have had an enormous and persistent influence on the popular image of the Columbian voyages and the native Americans.

This, the second Basel edition of the *First Letter from America,* was the first book to provide illustrations for the Columbus voyages. The printer Johan Bergmann de Olpe recycled illustrations from older publications. The book is open to an illustration of European wharves, which the printer tried to pass off as an American scene by labeling it 'Island of Española.'

5.
Cristoforo Colombo, 1451–1506. *Eyn schön hübsch lesen von etlichen insslen die do in kurtzen zyten funden synd durch de künig von hispania . . .* Strasbourg: Bartholomäus Kistler, 1497.

This is the first German edition of the *First Letter from America.* Typical of German printed books, it bears a beautiful woodcut illustration on the frontispiece. It is the first printed work to introduce a religious motif in the American story; Christ addresses King Fernando and his courtiers.

Spanish publishers knew that the Columbus voyage was a Castilian enterprise, and that Queen Isabel of Castile was the ruler of the new islands. But publishers in other parts of Europe could not seem to understand that Columbus reported principally to the queen, not the king.

6.

Fernando V, 1452–1516, and Isabel I, 1451–1504. *Libro en q[ue] esta[n] copiladas algunas bullas de n[uest]ro muy s[an]cto padre co[n]cedidas en fauor de la iurisdicio[n] real de sus altezas, & todas las pragmaticas q[ue] esta[n] fechas para la buena governacion reyno.* Alcalá de Henares: Lançalao Polono, 16 November 1503.

Having acquired formal recognition of Castilian sovereignty over La Española, Fernando and Isabel authorized colonization of the island. A colony would constitute effective occupation—one of the most important proofs of sovereignty. Events on La Española from 1493 through 1497, however, punctured the euphoria with which Spanish colonists had volunteered for the second voyage. The fleet made a fast safe crossing, but found the thirty-nine men left behind in La Navidad had been killed either by disease or by the local Indians. The natives that Columbus had described as meek and gentle in the *First Letter from America* turned out to be hostile and dangerous.

The environment also proved dangerous. More than three hundred Europeans became seriously ill after they went ashore, including Columbus, and many died. Columbus diverted his energies into conquering the Indians and exploring further west, where he still expected to find the Asian continent. He left the administration of the colony in the hands of his brothers Bartolomé and Diego Colón, who had just arrived from France and Italy, never having lived in Spain.

The Spaniards who had come to colonize the island became dissatisfied with the Columbus brothers' authoritarian and uninformed rule. Groups of colonists moved away from Columbus's headquarters city of La Isabela to form squatter settlements, but the Columbus brothers attacked them and arrested them as rebels. The colonists felt that the denial of their rights as Spanish citizens had become intolerable. The appellate judge Columbus had appointed for the whole island, Francisco Roldán, led a revolt in which almost one hundred colonists moved to the Xaraguá Peninsula and established a new town. Most of the other colonists simply gave up; they returned to Spain on the four resupply fleets that crossed between Spain and Española. The number of colonists remaining on the island fell below three hundred.

As all of this bad news trickled back with the returning fleets, the reputation of Columbus's colony deteriorated. When Columbus returned to Spain in 1497 and started organizing his third voyage, he could not recruit enough Spaniards to man his ships or settle his colonies.

In desperation, he asked the monarchs for permission to recruit condemned murderers, who would gain commutation of their death sentences after ten years of unpaid labor in the colonies. This royal provision of 1497, as well as the 1493 papal decrees establishing the Line of Demarcation, were quickly incorporated into books of Castilian laws such as this one.

The book is open to the royal permission for Columbus to recruit convicts. The text notes that this decree was announced by the public crier of the city of Granada, who read it aloud in the city's principal marketplace on October 14, 1497. Despite the inducement offered, Columbus was able to recruit only ten convicts, six of them Gypsies convicted of murder.

Columbus's third voyage was the first to carry European women to America. We know the names and identities of only four of these pioneer women: Catalina de Sevilla was accompanying her husband, noncommissioned officer Pedro de Salamanca; Gracia de Segovia appears to have been single and traveling alone; Catalina de Egipcio and María de Egipcio were Gypsy women convicted of murder who took advantage of the offer of commutation of sentence after ten years of unpaid service in the Americas.

7.
Amerigo Vespucci, 1454–1512. *Alberic' vespucci' laure[n]tio petri francisci de medicis salutem plurima[m] dicit.* Paris: Félix Baligaut and Jean Lambert, 1503.

Columbus explored the coast of Venezuela in 1498. Realizing from the volume of water pouring out of the Orinoco River delta that this was a continent, he informed the monarchs that he had taken possession of this "Other World" in their name. Fernando and Isabel sent Alonso de Ojeda and Juan de La Cosa the next year to establish colonies on the mainland of South America and explore further for a passage to Asia. The Florentine merchant, Amerigo Vespucci, went on this voyage, probably as agent of the Sevillian merchants who had invested in the costs of the voyage.

After he returned to Spain in June 1500, Vespucci sent this letter to Lorenzo Pier Francesco de' Medici, cousin of Lorenzo de' Medici the Magnificent, describing the voyage. In Florence, Giovanni Giocondo translated the letter into Latin.

In this letter, Vespucci claimed that he had made astronomic observations and measured longitude by a new method, using the conjunction of the moon with Mars on August 23, 1499, as his reference. Scholars now doubt that he made observations of lunar distances off the Venezuelan coast.

8.
Pietro Martire d'Anghiera, ca. 1457–1526. *Libretto de tutta la nauigatione de re de Spagna de la isole et terreni nouamente trouuti.* Venice: Albertinus Vercellensis, 1504.

As Spanish fleets continued to explore and map the American coastline, King Fernando and Queen Isabel drew on every possible resource to publicize their sovereignty. One of the most effective publicists was an Italian humanist, Pietro Martire d'Anghiera, who had been a chaplain on Isabel's staff since 1488. Assigned responsibility for writing the history of the explorations, d'Anghiera chose to do so in the form of letters to friends and patrons in Spain and Italy. These letters shaped European impressions of the Americas.

The popularity of the letters stemmed from a paradox; using the classical style admired in intellectual circles, they narrated high adventure in exotic places. D'Anghiera never saw the Americas, though he was appointed abbot of the island of Jamaica. He acquired his information in the traditional way of Castilian royal chroniclers, by interviewing the participants (he was acquainted with Columbus, Cortés, Magellan, Cabot, and Vespucci), or reading the written reports captains submitted to the House of Trade in Seville. D'Anghiera shaped these first-hand accounts into an elegant narrative true to the norms of classical rhetoric. His first collection of letters, written between 1488 and 1525, recounts the first three voyages of Columbus. With d'Anghiera's permission, the letters were translated into the Venetian dialect by Angelo Trevisan, Venetian ambassador to Spain, with the translator's description of Columbus at the beginning of the collection.

The *Libretto* is the earliest known publication of Pietro Martire d'Anghiera, the first historian of the Americas. It is the first published collection of his letters; it contains the first account in printed form of the third voyage and a part of the second voyage of Columbus. This beautiful copy, another in the John Carter Brown Library, and a third lacking the title leaf in the Marciana Library in Venice are the only recorded surviving copies.

The provenance of this copy of the *Libretto* is faithfully and charmingly related by Douglas G. Parsonage:

> On a certain visit to Budapest in 1929 Lathrop C. Harper called as usual on his old friend Gustav Ranschburg, antiquarian bookseller. Civilities were exchanged, companionable gossip started, and then the inevitable request from Harper: "Let's see some books." I was not there myself and cannot fill in the details as to what was shown and rejected, what else was looked at and set aside for further consideration. But I do know that at a certain point a nondescript, unimpressive volume in vellum lettered on the spine *Opusc. Var.* came to light. It was a volume containing twelve sixteenth-century pamphlets bound together with a manuscript index, with such unpromising titles as *Conjectura de libris de imitatione Christi; Dissertatio de Tracala; De numismatis et nummis antiquis;* and others.
>
> Knowing Mr. Harper, I would like to imagine a certain bird-dog attitude that may suddenly have developed. On the other hand, his poker face when the stakes were high was notorious. But I only have his own version of what happened next, and that was a laconic: "I asked him the price, and said I'd take it." One has to remember that all the titles were in Latin, of which Mr. Harper had little or no real knowledge; that the first nine pamphlets could have no possible interest for Harper or his customers; that the tenth was listed in the manuscript index as *Navigazione alla scoperta d'America,* Venezia, 1504; and the last two were similar to the first nine. Yet I have Mr. Harper's word, which I know to be true, that this is the title that caught his eye.
>
> The book itself, under its proper title of *Libretto de Tutta la Nauigatione de Re de Spagna . . .* is the first printed collection of voyages, derived from Peter Martyr and covering the first three voyages of Columbus. In his voluminous reading Harper must somewhere have encountered the title and remembered it. No copy has ever appeared for sale, and it is so rare that few bibliographers have ever seen it or referred to it, except to mention its legendary rarity. And here was a second perfect copy, nestling anonymously in the middle of a dull collection of tracts in a utilitarian vellum binding lettered *Opusc. Var.*!
>
> It came to America in Harper's pocket; was separated and placed in a morocco case, and kept side by side with the parent volume in Harper's safe until he died. To my knowledge he never offered it for sale, and indeed few people knew he had it. During the thirties times were bad, and few libraries or collectors had the kind

of money Harper felt it was worth—and rightly! Came the forties and his slow withdrawal from active bookselling, partly due to advancing years and partly to family inheritance matters that occupied most of his time, and at his death in 1950 it was still in the safe.

With the sale of the business after Harper's death, and the availability of a typed inventory of the books in his stock, some of the more knowledgeable dealers tried to buy the business for this one book. But only one of several collectors who saw the list was as alert as Harper had been to this outstanding rarity, which was typed out on a single line without any comment or price with the four or five thousand other books in the Harper stock. That Bernardo Mendel was the eventual successful purchaser of the business against considerable odds is evidence of the knowledge and acumen that went into forming his own library that now enriches the shelves of the Lilly Library at Indiana University. This book was a gift of Mr. Mendel on the occasion of the dedication.

The book is open to Trevisan's description of the personal appearance of Columbus:

Christopher Columbus, Genoese, a man of tall and imposing stature, ruddy-complexioned, of great intelligence, and with a long face, followed the Most Serene Sovereigns of Spain for a long time wherever they went.

9.
Amerigo Vespucci, 1451–1512. *Mundus Novus.* Augsburg: Johann Otmar, 1504.

Amerigo Vespucci returned from Ojeda's explorations and was summoned to Lisbon by King Manuel of Portugal. He left Seville so hastily that he did not have a chance to say goodbye or gather his belongings. The king, after receiving news of Pedro Alvares Cabral's accidental landfall on Brazil (April 23, 1500), ordered follow-up explorations. Probably, King Manuel, having learned that Amerigo had found a new system for measuring longitude, wanted him to fix the position of Brazil with respect to the Line of Demarcation to see whether it was east or west of the line. If it was west, it would belong not to Portugal but to Spain.

On May 13, 1501, Amerigo sailed with a fleet of ships in the service of Portugal. They reached the South American coast at Cape Saint Augustine (8°64'S), went south to latitude 46'S, and returned to Lisbon in July 1502. Amerigo wrote letters to his Florentine patron describing this, his second and last voyage to the Americas.

The book is open to the title page, where Amerigo calls the Americas "A New World." In this letter, as in his others, Amerigo gives the false impression that he commanded or piloted a ship, never mentioning Ojeda or any other captain.

10.
Amerigo Vespucci, 1451–1512. *Be [i.e. De] ora antarctica per regem Portugalli pridem inuenta.* Strasbourg: Matthias Hupfuff, 1505.

In 1505, the German humanist Matthias Ringmann translated and published Amerigo Vespucci's second letter to Lorenzo Pier Francesco de' Medici, in which he described his voyage with the Portuguese to South America.   Ringmann changed its title from *New World* to *About the Antarctic coast long ago discovered through the king of Portugal.*

The book is open to an illustration that purports to show Brazilian natives. This picture had a powerful influence on Europe's image of native Americans. In fact, the illustration was drawn by a Strasbourg artist who never saw a native American; he based his drawing on Renaissance concepts of the ideal human body.

11.
Martin Waldseemüller, 1470–1521? *Cosmographiae introductio cum quibusdam geometriae ac astronomiae principiis ad eam rem necessariis . . .* Saint-Dié: Gautier Lud, 1507.

This is the book that gave the name "America" to the newly discovered lands in the West.

In 1506, Martin Waldseemüller and several other humanists were working together at the court of Duke René, in the little town of Saint-Dié, in the Vosges Mountains of northeastern France. The humanists and their patron shared an interest in astronomy and cosmography. Together they were compiling a world atlas based on Ptolemy's *Geography.*

The German humanist Matthias Ringmann joined the Saint-Dié group, bringing his translation of Vespucci's second letter. In comparing Vespucci's account to the works of Ptolemy, he discovered that the lands Vespucci described related to a new world not mentioned by Ptolemy, a world apparently lying under the Antarctic pole. Excited by this discovery, the humanists decided to produce a world map to include the lands not mentioned by Ptolemy.

Waldseemüller and Ringmann wrote this small tract on cosmography to serve as an introduction to the world map as well as to their planned atlas. It was entitled *An introduction to cosmography with several elements of geometry and astronomy required for this science, and the four voyages of Amerigo Vespucci.* In the text the authors suggested that the new lands, which formed the fourth part of the world, should be named "Amerigo" or "America" to honor Vespucci, whom they believed had discovered them. The little volume was widely distributed.

A few years later, Waldseemüller realized that these were the same lands earlier discovered by Columbus on his third voyage. He removed the name "America" in later editions, but by then it had become generally accepted. Despite all the efforts by Columbus and the Spanish monarchs to publicize Castile's discoveries and claims, the name of the Western Hemisphere would forever belong to Vespucci, an Italian businessman who had no role in its first exploration by Europeans.

The book is open to the Saint-Dié humanists' introduction to Vespucci's letter, where the authors give their reason for the name "America":

> Now, really these [three] parts [Europe, Africa, and Asia] were more widely traveled, and another fourth part was discovered by Americus Vesputius (as will be seen in the following pages), for which reason I do not see why anyone would reasonably object to calling it (after the discoverer Americus, a man of wisdom and expertise) "Amerige," that is, land of Americus, or "America," since both "Europa" and "Asia" are names derived from women.

12.

Hernando Cortés, 1485–1547. *Praeclara Ferdinando Cortesii de noua maris oceani Hyspania narratio . . .* Nuremberg: Friedrich Peypus, 1524.

By 1515, Spaniards had established twenty-seven cities and towns on the four major Caribbean Islands—Española, Cuba, Jamaica, and Puerto Rico. They had tried to establish cities and towns on the mainland, the east coast of which they had explored and mapped from Brazil to southern Georgia. Yet every Spanish colony on the mainland had been destroyed by epidemic, crop failure, or native attack. If they were going to secure sovereignty over the mainland, they would have to conquer the native rulers first. This means of effective occupation became the norm in the European colonization of the Americas.

Hernando Cortés led the first successful conquest of a native ruler on the American mainland. Leading an expedition of about four hundred men, Cortés sailed from Cuba in 1518 and won the support of dozens of Indian city-states that had been conquered by Moctezuma's expanding Mexica (Aztec) empire. With the assistance of about 100,000 native warriors from these allies he had conquered Moctezuma's capital city of Tenochtitlán (Mexico City) by 1522.

Cortés described his expedition's progress, defeats, and victories in five long letters he sent to the new ruler of Spain, Charles V. In his second letter, Cortés described the allied army's march from the coast at Veracruz to the valley of Mexico, Moctezuma's capitulation to the invading forces, and the allies' occupation of the capital. The Spaniards, who considered cities to be the mark of a civilized society, marveled at the sophistication and monumentality of Mexico City. They also expressed horror and disgust at some of the Mexica religious rites.

The book is open to a map of Mexico City that accompanied Cortés's second letter to Charles V. This is the first map of any American city. This map is also the first to contain the designation "La Florida." It lies to the northeast of the Yucatán, Cuba, and the Panuco River.

Within the city, we can see its canals and massive architecture, including pyramids and ball court. We can also see the skull rack, where the Aztecs displayed the heads of captives and hostages who had been sacrificed in the temple of the main pyramid. In this Latin translation, the skull rack is labeled *capita sacrificatorum.*

13.

Escrituras casa en la calle real, Mexico. 1527, May 9–1648, July 13. Latin American mss. Mexico.

Although the Mexica succeeded in defeating and expelling the Spaniards and their allies from Tenochtitlán in 1520, Cortés retreated to the allied city of Tlascala and formed a new army with reinforcements from another Spanish expedition. This much larger force definitively conquered the Mexica's new lord Cuauhtemoc, who had led a heroic resistance.

Immediately after the conquest, the Spaniards occupied the land and buildings. In order to distribute the property according to Castilian law, Cortés carried out a survey of all the property in the city. He took the property, tribute, and other income that had belonged to the ruling dynasty and turned it over to the Spanish monarchy. The properties and income of the native temples became the property of the church. Private houses and lots in the city became the booty of the conquerors.

This manuscript is a bill of sale for a house and its lot in Mexico City. Cortés had given the lot to one of the conquerors, Juan Ximénez, in 1524, and on May 9, 1527, Ximénez sold it to another Spaniard, Ruy Pérez, for 100.5 pesos of gold.

The manuscript is open to the 1527 bill of sale containing the property description:

> A lot that I have and own in this city [of Temistitan], which is adjoined on one side by the property of Alonso Galeote, on the front by the street, and in back by the property of Hernando de Xerez.

The sale included the lot, buildings, and some building stone.

14.

Antonio Pigafetta, ca. 1480/91–ca. 1534, and Transylvanus Maximilianus. *Il viaggio fatto da gli Spaniuoli a torno a' l mondo . . .* Venice: Lucantonio Giunta, 1536.

In 1519, an Italian merchant from Vicenza, Antonio Pigafetta, booked passage on a fateful voyage departing from the Spanish port of Sanlúcar de Barrameda, bound for Asia. Portuguese fleets had been traveling around Africa to Asia for ten years, but this expedition was different; the Portuguese captain of this fleet of five ships, Ferdinand Magellan, was commissioned by Spain to find a route through or around South America in order to cross the Pacific and reach Asia.

Spaniards had no illusions that this would be an easy voyage. In 1515, Castile had sent an expedition with this same objective under the command of the royal pilot Juan Díaz de Solís. The Solís expedition, after carefully exploring the coast of South America from Brazil to Argentina, entered the Río de la Plata, where Solís was killed by natives, and the expedition fell apart. Where Solís had failed, Magellan succeeded, though he too was killed by natives, in the Philippines, before completing his voyage.

At the end, a Spanish seaman named Juan Sebastián de Elcano brought the fleet back from Asia. In the Spice Islands they sold the leaking *Concepción* to buy a cargo of cloves, but *Trinidad*'s captain and crew chose to go back across the Pacific rather than continue west. Elcano navigated the lone surviving ship *Victoria* around Africa, and arrived home in September 1522, three years after the expedition had departed. *Victoria*'s cargo of cloves fetched such a high price on the market that it paid for all the expedition's expenses and turned a handsome profit for the expedition's stockholders.

The news of this first circumnavigation of the globe spread rapidly. Hernando Cortés heard it the same year from the captain of a Spanish colonizing expedition he ran into on the coast of Honduras; he immediately wrote a note of congratulations to Emperor Charles V. Yet for all the significance of this, Spain's first voyage to the Spice Islands, we would know very little were it not for the Italian passenger. Pigafetta wrote an account of the voyage in which he chronicled the terrors of the violent passage through the Straits of Magellan, the horrors of starvation and lack of drinking water during the Pacific crossing, and the many intrigues and betrayals once the fleet reached Asia.

This is the first Italian translation of Pigafetta's narrative. The book is open to the final pages, which give a brief Brazilian vocabulary.

15.
Charles V, Holy Roman Emperor, 1500–1558. *Letter to Antonio de Mendoza.* 1537, Feb 16. Latin American mss. Mexico.

As Spanish explorers penetrated the interior of Mexico and moved farther north, the monarchy delegated to the viceroy of New Spain the power to draw up contracts, or capitulations, between the crown and the explorer. The first viceroy of New Spain, Antonio de Mendoza, arrived at his post in Mexico City in 1535 and immediately began regularizing the government and extending Spain's effective occupation of the North American continent.

A couple of years later, Emperor Charles V sent him this letter recommending Francisco Vásquez Coronado (1510–1554). The king reminded his viceroy that Coronado's father and brother had both served the Spanish monarchy and suggested that Coronado be given important assignments that would benefit the monarchy.

Clearly, the king was approving an appointment that Mendoza had already suggested. In 1539, the viceroy commissioned Coronado as commander of an expedition that was being assembled to explore the present southwest United States. The Coronado expedition, comprising 336 Spaniards, hundreds of Indian guides and bearers, and more than 1,500 horses, mules, and livestock, began moving north from Culiacán on April 22, 1540.

During the next two years, the Coronado expedition explored Arizona, New Mexico, Texas, Oklahoma, and Kansas. Without being aware of it, the expedition at one time was within seven hundred miles of the De Soto expedition, which was exploring westward from Florida and Georgia to the

Mississippi. The Coronado expedition began its return march to Mexico City in early April 1542, after Coronado had been seriously injured in a riding accident. For Europeans, his explorations transformed North America from an imaginary island into another new continent for colonization.

16.

Leyes y ordenanças nueuame[n]te hechas por su magestad, pa[ra] la gouernacion de las Indias y buen tratamiento y conseruacion de los Indios: que se han de guardar en el consejo y audie[n]cias reales q[ue] en ellas residen: y por todos los otros gouernadores juezes y personas particulares dellas. Alcalá de Henares: Juan de Brocar, 1543.

The laws that had traditionally governed Spain soon proved to be inadequate to deal with the many new situations that arose as Spaniards and native Americans struggled to live together. On the recommendation of many colonists and royal officials, Emperor Charles V issued new laws for the Americas. These laws were designed to improve the administration of the colonies and provide a more effective judicial system.

The most controversial parts of the New Laws attempted to improve the situation of the native Americans and eliminate colonists' control over Indian labor. Colonists who had the most to lose objected vehemently, so the viceroys implemented the laws gradually and piecemeal, compromising in some situations and discarding ordinances in others. By about 1570, three-quarters of the grants of Indian labor to private individuals had been converted to taxes and service paid to the royal treasury. By the end of the century, the old system of Indian labor was effectively ended. Indians in most of Spanish America had become, like Spaniards themselves, taxpayers to the royal government rather than forced laborers.

17.

Juan Ginés de Sepúlveda, 1490–1573. *Apologia Ioannis Genesii Sepulvedae pro libro de iustis belli causis ad amplissimum, & doctissimum praesulem.* Rome: Valerio and Luigi Dorici, 1550.

From the beginning of the American enterprise, many Spaniards were troubled by the moral and legal questions posed in the Americas. Particularly vexing was the question of whether native Americans should be allowed to continue their own life style or should be integrated into the mores of Europeans.

Dr. Juan Ginés de Sepúlveda, eminent humanist at the University of Salamanca and translator of Aristotle, used the reasoning of the ancient Greek philosopher to address this question. Employing Aristotle's theory of natural slavery, Sepúlveda argued that the natives' political inferiority, idolatries, and other sins were in their very nature. They were not capable of change by their own volition or by the example of Europeans. Only by war and consequent enslavement could barbarous customs such as idolatry, cannibalism, and human sacrifice be ended.

Sepúlveda also based his argument on the papal decree of 1493 that divided the Spanish and Portuguese spheres of exploration. Because the pope had granted sovereignty over the Western Hemisphere to Spain and charged the Spanish monarchy with responsibility for converting the natives to Christianity, Spain should end these practices abhorrent to Christians. Conquest of the Indians was necessary, and therefore constituted a just war.

The book is open to Sepúlveda's discussion of the papal decree and its logical consequence, war against barbarous Indian practices in the Americas.

18.
Bartolomé de Las Casas, 1474–1566. *Tratado co[m]probatorio del Imperio soberano y principado universal que los reyes de Castilla y León tienen sobre las Indias.* Seville: Sebastián Trugillo, 1553.

The greatest opponent of Sepúlveda's conclusions, though not of his reasoning, was Bartolomé de Las Casas, a Dominican friar who had lived in the Indies, where he had inherited from his father a grant of Indian labor on the island of Cuba. Horrified by the consequences of the Indian labor grants, Las Casas devoted most of his life to saving the Indians from the Europeans. In order to develop his power of argument, he studied logic and philosophy at the University of Salamanca.

Vehemently opposed to forced conversion, and therefore to conquest, Las Casas argued that the Indians were in an early stage of development, just as Europeans had been millennia before. They were capable of acquiring European culture by example and therefore should live in settled communities in the presence of priests and other good Europeans.

In 1550 Emperor Charles V summoned jurists, including Sepúlveda and Las Casas to debate the issue. The famous debate before the emperor in Valladolid produced no resolution. It is important to remember that Las Casas and Sepúlveda both wanted to achieve the same objective—conversion of the Indians to Christianity and the end of their barbarous practices—and that they both used Aristotelian logic to make their arguments.

The volume contains a collection of nine tracts by Las Casas. Here, it is open to the first tract, where, in order to lay the basis for the reforms he was demanding, Las Casas presents proofs that Spain has legitimate sovereignty by virtue of the papal decree of 1493.

19.
Philip II, King of Spain, 1527–1598. *Grant of liberty to the town of Tlascala.* 1556, August 24. Latin American mss. Mexico.

As the conquest generation aged and passed away, the lines between Indian and Spaniard became blurred. Spaniards and Indians married; their children were legally considered Spaniards, regardless of race or language. Spanish colonists and American Indians became bilingual, using both Spanish and the dominant local Indian language. Government policy integrated Indian communities into the Spanish political system.

Some Indian communities were particularly eager to obtain the benefits of full political integration. Hernando Cortés had been able to conquer the Aztec capital largely because he persuaded dozens of Indian towns to become his allies. Among the first and most important of these allies was the Indian town of Tlascala. After the conquest, the Spanish government tended to treat all Indian towns alike, despite their varying roles in the conquest. This, of course, offended and worried the Tlascalans. They sued, protesting that their original agreement had been violated.

In 1556, King Philip II granted this charter of liberty to the town of Tlascala. He assured the Tlascalans that they would forever enjoy all the freedoms, tax exemptions, and privileges of a Spanish town.

The document is open to the text issued by the royal chancery and its translator, with matching texts in Spanish and the Aztec language, Nahuatl.

20.
Títulos de las casas que compró Catalina Vázquez en el barrio de Tomatlán. 1562, Oct 4–1597, Nov 18. Latin American mss. Mexico.

Both Spaniards and Indians made full use of the judicial system to press their claims and ask for redress of grievances. In property disputes, the litigants submitted testimony and legal documents in both languages and also in the pictorial tradition of Nahuatl manuscripts.

This bill of sale is written in both Spanish and Nahuatl, and the property is sketched showing the Indian owners in residence. The owner of one house is don Diego de San Francisco and the other is doña María Francisca.

21.
Juana de Austria, 1535–1573. *V. Md. aprueba el assiento que se ha tomado con don Francisco de Mendoça.* 1558, Dec 10. Latin American mss. Mexico.

The original intention of Fernando and Isabel had been to find a western route to Asia and its spices. In the mid-sixteenth century, the Spanish monarchs were still trying to enter the spice trade. In Francisco de Mendoza they thought they had found the ideal spice producer. Mendoza had gone to New Spain (modern Mexico) as a teenager with his father, Antonio de Mendoza, the first viceroy. For twenty years, Francisco assisted his father and engaged in business.

After his father's death, Francisco returned to Spain. He was looking for a profitable investment for the silver he had brought back, both for himself and for other colonists who had made their fortunes in the Americas. The American investors thought that Francisco would be the ideal investment manager in Spain; his relatives were among the most powerful nobles at court.

In this royal concession, Princess Juana granted Francisco a monopoly on the planting and marketing of Asian spices—Chinese ginger and sandalwood—in New Spain. In return for land, a sufficient number of Indian laborers, and startup tax exemptions, Francisco agreed to pay the monarchy one-half of all the profits on his spice sales.

All of this came to nothing, however. A few weeks later, the princess commissioned Francisco to inspect all the mines in Spain and recommend which ones the monarchy should develop. Once Francisco had completed this assignment, the king appointed him director of the royal mines, and he spent the rest of his life in the royal service in Spain. He never went back to New Spain. Other Spaniards took up the project, however, and a century later ginger had become one of the leading cash crops of the town of Coamo in Puerto Rico.

The last page of the manuscript displays the signatures and rubrics of all the officials who signed off on this contract, including the princess and the king.

22.
Richard Hakluyt, 1552–1616. *The principall navigations, voiages and discoveries of the English nation, made by sea or ouer land . . . .* London: George Bishop and Ralph Newberie, 1589.

While the Spanish monarchy sponsored dozens of expeditions of exploration by sea and land and Spanish colonists founded and settled more than a hundred cities and towns in the Americas, the English made no effort to establish their presence in the New World. English fishing fleets, along with ships from Portugal, Spain, France, and Scandinavia, made seasonal trips to the fishing banks of North America, drying and salting their catch onshore and then carrying their cargo back to Europe in time for the Lenten season. The French king Francis I made several unsuccessful attempts to claim and settle parts of North America and Brazil, but in England neither the queen nor Parliament regarded the Americas as an important objective of royal policy; they were busy conquering and colonizing Ireland.

One Englishman and his son, both named Richard Hakluyt, had very different ideas. They had the mentality of the space race in modern times and began publicizing their views with Parliament and the royal court. They convinced Sir Humphrey Gilbert and, after Sir Humphrey's death, his half-brother Sir Walter Raleigh that England should challenge Spanish sovereignty over North America by establishing effective occupation. Sir Walter organized and financed a colonizing voyage, which established a short-lived settlement on Roanoke Island, Virginia.

Despite the failure of the Roanoke venture, Hakluyt remained convinced that the English should colonize North America. In order to publicize this idea and bring pressure on the royal government and Parliament, Hakluyt began publishing narratives of early voyages to North America. For the foreign voyages, he commissioned translations into English, but his principal objective was to use the early English voyages as a legal basis for English sovereignty in North America. Hakluyt's efforts bore fruit in the seventeenth century, when England was devastated by religious persecution and civil war. Hundreds of thousands of English people fleeing the turmoil and dangers of their own country turned to Hakluyt's narratives and, believing his rosy picture of America, emigrated to Virginia and New England.

Publication of these narratives of exploration in English translation continues to the present day, under the title *Hakluyt's Voyages.*

23.

Francisco Fernández de la Cueva Enríquez, d. 1733, *et al.* *Documents concerning the sending of twelve Indian families to Florida.* 1703, Sept 11–1704, Apr 5. Latin American mss. Mexico.

With English settlers moving into Virginia and Georgia and making incursions into Florida, the Spanish government wanted to encourage Spanish colonies in North America. Those in Florida were particularly important because they secured the coast along the major route for the return voyage of the annual silver fleets.

The government decided to use a method that had developed in other frontier colonizations. Traditionally, Spaniards and Indians from already-colonized regions went to settle the northern frontier. The colonists who settled New Mexico in the late sixteenth century, for example, came mostly from the silver mining towns of northern Mexico.

This time, the viceroy of New Spain, the duke of Albuquerque, found Indians from New Spain who were willing to emigrate to Florida. The manuscript is open to the first page of the viceroy's decree authorizing twelve Tlascalan families to colonize Florida. The signature reads "Duque de Albuquerque."

24.

Ulrich von Hutten, 1488–1523. *Of the wood called guaiacum, that healeth the frenche pockes, and also helpeth the goute in the feete, the stone, palsey, lepre, dropsy, fallynge euyll, and other diseses. Made in Latyn by Ulrich Hutten, knyght, and translated in to Englysh by Thomas Paynel.* London: Thomas Berthelet, 1540.

With the Atlantic transformed from obstacle to conduit, people, plants, and animals moved back and forth between Europe and America for the first time. Species that had existed only on one side of the Atlantic now became common on both sides, a process we now call the Columbian Exchange.

One of the negative effects of the Columbian Exchange devastated native American cultures; disease-bearing microorganisms from Europe infected hundreds of thousands of American Indians. Diseases such as measles, typhus, and small pox had not existed in the Americas, and when the indigenous population was first exposed, they all got sick at once. When most of a community is seriously ill at the same time, there is no one to care for the sick. People who might have survived with adequate drinking water, food, and clean conditions died of dehydration and secondary infections.

Only one disease, syphilis, traveled in the opposite direction. There now seems little doubt that syphilis was in the Americas prior to 1492 and came to Europe on Columbus's return in 1493. Most Europeans called it the "French disease," because the first cases appeared in the French army that invaded Italy in 1494.

By 1500, syphilis had become an epidemic, spreading rapidly throughout Europe in a pattern characteristic of a new venereal disease in a population having no prior immunity.

Diagnoses and treatments were suggested by many physicians, including the Italian Girolamo Fracastoro and the Spaniard Pedro Mexía (who called it the "Indies disease"). The afflicted desperately sought any concoction that might give relief from the painful and disfiguring symptoms, even when the treatment was as bad as the disease. A German knight who was infected with the disease, Ulrich von Hutten, published this treatise advocating the use of guaiacum, the extract of an American wood. The book was immediately popular throughout Europe and appeared in English translation in 1540.

25.

Nicolás Monardes, ca. 1512–1588. *Primera y segunda y tercera partes de la historia medicinal de las cosas que se traen de nuestras Indias Occidentales que sirven en medicina.* Seville: Alonso Escrivano, 1574.

Physicians before the twentieth century had few sources of medication at their disposal. Most of the remedies they supplied to their patients were powders and syrups extracted from common plants, such as herbs and bark, and from minerals. When Columbus and other explorers brought back American species, European physicians eagerly acquired samples and asked how native American healers had traditionally used them. By the last quarter of the sixteenth century, Spanish physicians had accumulated years of experience with these American herbs.

Nicolás Monardes received his doctorate in medicine from the University of Seville and practiced medicine in that city, where he also owned a business importing drugs from the Americas. Some of his seven children went to America, but Monardes learned about American drugs from the cargo brought to Seville.

Monardes became famous throughout Europe after publishing this book on American drugs. Monardes was an expert botanist and illustrator. Because of his careful descriptions of drugs and the tests he carried out in animals to ascertain their medicinal properties, he is considered one of the founders of experimental pharmacology. He became the best known and most widely read Spanish physician in Europe in the sixteenth century.

The book is open to the chapter in which Monardes gives instructions on how to prepare and prescribe "Michoacan root."

26.

Nicolás Monardes, ca. 1512–1588. *De simplicibus medicamentis ex occidentali India delatis, quorum in medicina usus est . . .* Antwerp: Christophe Plantin, 1574.

This Antwerp edition of Monardes's book on the medicinal uses of American plants is open to his discussion of tobacco as a medicine.

27.

Nicolás Monardes, ca. 1512–1588. *Delle cose che vengono portate dall'Indie occidentali pertinenti al'uso della medicina.* Venice: Giordan Ziletti, 1582.

Monardes provided numerous illustrations of American plants he recommended as medicines. These drawings made the book so popular that it went through at least fifty editions in many languages. The English translation bore the wonderful title *Joyfull newes out of the newe founde worlde.*

28.

Gonzalo Fernández de Oviedo y Valdés, 1478–1557. *La historia general de las Indias.* Seville: Juan Cromberger, 1535.

In the first century of colonization, Europeans who stayed at home wanted to know the ways that native American societies were different from their own. Consequently, colonial writers filled their books with descriptions and illustrations of every aspect of native America that was unknown in Europe. When colonial writers did not describe some aspect of Indian society, we can safely assume that it was considered too similar to Europe to warrant mention.

Gonzalo Fernández de Oviedo spent most of his life in the Americas as a royal inspector of mines and as commander of the fortress at Santo Domingo. Fascinated by the local customs and the exploits of the Spanish conquerors, Oviedo recorded his observations of native customs, artifacts, flora, fauna, and language, and he interviewed participants in Spanish expeditions to the American mainlands.

An incurable reviser of his own work, even after publication, he composed, expanded, and rewrote his books on board ship during his thirteen transatlantic voyages. To him we owe much of what we know of Indian words, customs, and skills.

In his second book, *On the general and natural history of the Indies,* Oviedo described how the islanders made their dugout canoes, even though they had no metal tools to work with:

> On this island of Española and in all other coasts and rivers of the Indies that Christians have seen up to now, there is one type of boat, which the Indians call 'canoe.' With these they navigate the great rivers as well as the oceans for purposes of war and raid, for conducting commerce between one island and another, or for fishing and other purposes. Without these canoes, we Christians who live here could not benefit from our property on the coasts and riverbanks.

> Each canoe is a single branch or tree trunk, which the Indians hollow with blows of stone-headed axes, like the one illustrated here. With these they cut or scrape the wood to hollow it. Then they burn what has been beaten and cut little by little. When the fire goes out, they again cut and beat it, alternating these two processes until the boat is formed to the limits of the width and length of the tree trunk.

I've seen them large enough to carry forty-five men, wide enough to hold a wine cask easily between the Carib Indian archers. The Caribs use them as large as I have mentioned and call them pirogues, navigating with cotton sails and by oar, as well as with their paddles, as they call their oars. Sometimes they paddle standing, at times sitting, and kneeling when they feel like it.

Some of these canoes are so small that they hold no more than two or three Indians, others hold six, others ten, and on up. But no matter what size, they are very light and dangerous, for they overturn frequently. But they don't sink even when they are immersed. And because these Indians are great swimmers, they turn them upright and smartly empty them . . . . They are safer than our ships in case of capsizing. Even though ours capsize less often because they are more flexible and buoyant, once they are immersed they sink to the bottom.

29.

Gonzalo Fernández de Oviedo y Valdés, 1478–1557. *Coronica de las Indias.* Salamanca: Juan de Junta, 1547.

The book is open to an illustration of two American novelties: an Indian drum and a tobacco pipe.

30.

Gonzalo Fernández de Oviedo y Valdés, 1478–1557. *L'histoire naturelle et generalle des Indes, isles, et terre ferme de la grand mer oceane.* Paris: Michel de Vascosan, 1555.

Oviedo was fascinated by the techniques American Indians had developed for familiar tasks, such as preparing food, building houses, and manufacturing tools and equipment. Here he illustrates the Indian method of starting a fire without the flint and steel that Europeans required. The native Americans rotated a stick poised on a piece of wood by rubbing the stick between their hands to create heat.

31.

Denis le Chartreux. *Este es un compendio breue que tracta d'la manera de como se han de hazer las processiones.* Mexico City: Juan Cromberger, 1544.

In 1539, the first printing press arrived in the Americas. It was established in Mexico City as a branch of the Seville press owned by the Cromberger family. Juan Cromberger in Mexico followed the same policies he had learned in the family firm in Seville. In Europe, printing had been a business from the beginning. Most printers were motivated by profit and they tended to reproduce those texts that had already been most in demand before the advent of printing.

Denis the Carthusian's little book on how to lead a Christian life had been a Latin classic for two centuries when the new bishop of Mexico, Juan de

Zumárraga, commissioned this printing of Denis's section on processions, in Spanish translation.

Clearly, the good bishop thought that Spanish colonists in America were behaving no better than their medieval predecessors in Europe. The section that he commissioned from Cromberger points out the behaviors that are not suitable for holy processions: foolishness, pompous attire, drinks inappropriate to the time and place, and above all buffoonery and other lewd acts that provoke carnal sins.

32.

Francisco López de Gómara, 1510–1560? *La istoria de las Indias y conquista de Mexico.* Saragossa: Agustín Millán, 1552.

The secretary of Hernando Cortés, Francisco López de Gómara, never saw America, yet he made excellent use of reports and descriptions from members of the Cortés expedition. Gómara devotes three chapters to Aztec women, focusing on practices that surely would arouse the curiosity of Europeans; how women are buried, the practice of polygamy, and marriage rites.

The book is open to a description of the Aztec writing and number systems and the names of the months and days in the Aztec calendar.

33.

Pedro de Cieza de León, 1518–1560. *Parte primera de la chronica del Peru.* Seville: Martin de Montesdoca, 1553.

While Spaniards regarded most Indian customs with simple curiosity, other practices aroused disgust and horror. The most abhorrent were human sacrifice and cannibalism, both practiced in Mesoamerica and Peru.

The chronicler of the conquest of Peru, Pedro de Cieza de León, admired much of Inca society and political organization. But in a chapter entitled "On the rites and sacrifices of those Indians and what great butchers they are for eating human flesh," he roundly condemned the Inca practice of human sacrifice: "Judging by the sins of these people, the devil, enemy of human nature, wields enormous power and dominion over them."

Cieza de León illustrated this harsh judgment with a drawing of two human corpses hanging like sides of beef. In the foreground, a priest cuts out the heart of a captive, while in the background the devil dances atop a column symbolizing justice.

34.

Hans Staden. *Warhaftig Historia un Beschreibung eyner Landtschafft der Wilden.* Marburg: Andreas Kolbe, 1557.

The most sensational book about the American Indians was a narrative written by a young German adventurer, Hans Staden. Staden signed on as a gunner for a Portuguese voyage to Brazil and was captured by a cannibal tribe, the Tupinamba. How he managed to stay alive for several years while his fellow crew members were killed and eaten is a story that takes up most of his narrative. His book became a sensation and deeply influenced European perceptions of Brazil, largely because the German publisher provided detailed and gory illustrations of how the Indians killed and barbecued their captives.

35.
Domingo de Santo Tomás, 1499–1570. *Grammatica, ò arte de la lengua general de los indios de los reynos del Peru.* Valladolid: Francisco Fernández de Córdova, 1560.

While the sensationalist press continued to publish extravagantly illustrated accounts of American Indian sacrificial practices, more sober writers studied the fundamental aspects of native American culture.

Domingo de Santo Tomás was one of the most respected observers of Incan society. He wrote this, the first grammar of the Inca language, Kechua. The book is open to a page of Kechua vocabulary.

36.
Doctrina Christiana y catecismo para instruccion de los indios. Lima: Antonio Ricardo, 1584.

In contrast to the earlier Cromberger book in Mexico City, which was aimed at laymen, this little book was intended as a manual for priests. The Spanish priests in Peru already knew how to speak the native languages. In 1583, they came from all over Peru, meeting in the Spanish capital, Ciudad de los Reyes [Lima]. They translated the Catholic catechism into the two principal native languages of Peru, Kechua, and Aymara, and commissioned this printing from the press of Antonio Ricardo.

The catechism had been invented by Martin Luther, but Catholics soon compiled their own. In Europe, children were required to memorize the fundamentals of the Christian faith, in the form of prayers, the creed, and question-and-answer sequences. The book is open to the first prayers that every Catholic child learned, the "Lord's Prayer" and "Hail Mary." Here the prayers in Spanish are followed by their translations into Kechua in the left column and Aymara in the right column.

37.
Giovanni Battista Ramusio, 1485–1557. *Terzo volume delle navigationi et viaggi.* Venice: Giunti, 1565.

In North America, French explorers advanced up the St. Lawrence River in pursuit of fur pelts. Their numbers were few and their conflicts with the local Indians proportionately minor, compared to the Spanish expeditions to the south. The book is open to an illustration showing Indians near Montreal defending their village against the French. The artist has carefully displayed the construction of the palisade, the layout and construction of the dwellings, how the Indians threw rocks at the attackers, and American animals such as elk and polar bear.

38.
Jean de Léry, 1534-1611. *Historia navigationis in Brasiliam quae et America dicitur.* Geneva: Heirs of Eustache Vignon, 1594.

Jean de Léry recorded the history of an attempted French settlement in Brazil in the mid-sixteenth century. The settlement failed largely because of conflict among the colonists and hostility of the native Americans. Léry found much to comment on in the exotic flora and fauna. He provided careful drawings of Indian customs and regalia, and the publisher's woodblock cutter transformed the Indians into classic beauties in the poses of Greek gods. Here, he illustrates Indians dancing in feather headdress, using a gourd shaker, with a monkey in the foreground and a parrot in the background. On the facing page, Léry provides an example of an Indian chant.

¶ Epistola Christofori Colom ; cui etas nostra multū debet: de
Insulis Indię supra Gangem nuper inuentis. Ad quas perqui/
rendas octauo antea mense auspicijs ⁊ ere inuictissimi Fernan
di Hispaniarum Regis missus fuerat:ad Magnificum dñm Ra
phaelem Sanxis:eiusdem serenissimi Regis Tesaurariū missa:
quam nobilis ac litteratus vir Aliander de Cosco ab Hispano
ideomate in latinum conuertit : tertio kals Maij. M.cccc.xciij.
Pontificatus Alexandri Sexti Anno Primo.

Quoniam susceptę prouintię rem perfectam me cōsecutum
fuisse gratum tibi fore scio: has constitui exarare: quę te
vniuscuiusꝗrei in hoc nostro itinere gestę inuentęꝗ ad/
moneant: Tricesimotertio die postꝗ Gadibus discessi in mare
Indicū perueni:vbi plurimas insulas innumeris habitatas ho/
minibus repperi:quarum omnium pro foelicissimo Rege nostro
preconio celebrato ⁊ vexillis extensis contradicente nemine pos/
sessionem accepi:primęꝗ earum diui Saluatoris nomen impo/
sui:cuius fretus auxilio tam ad hanc:ꝗ ad ceteras alias perue/
nimus. Eam ꝟo Jndi Guanahanin vocant. Aliarum etiā vnam
quanꝗ nouo nomine nuncupaui. Quippe alia insulam Sanctę
Marię Conceptionis aliam Fernandinam . aliam Hysabellam.
aliam Johanam ⁊ sic de reliquis appellari iussi. Quamprimum
in eam insulam quā dudum Johanā vocari dixi appulimus: iu
xta eius littus occidentem versus aliquantulum processi:tamꝗ
eam magnā nullo reperto fine inueni:vt non insulam: sed conti
nentem Chatai prouinciam esse crediderim:nulla tñ videns op/
pida municipiaue in maritimis sita confinib?preter aliquos vi
cos ⁊ predia rustica:cum quoꝝ incolis loqui nequibam quare si
mul ac nos videbant surripiebant fugam. Progrediebar vltra:
existimans aliquā me vrbem villasue inuenturum. Deniꝗ vidēs
ꝗ longe admodum progressis nihil noui emergebat:⁊ hmõi via
nos ad Septentrionem deferebat:ꝗ ipse fugere exoptabā:terris
etenim regnabat bruma: ad Austrumꝗ erat in voto cōtendere:

1. Colombo. *First Letter from America.*

4. Verardi. *The Island of Española.*

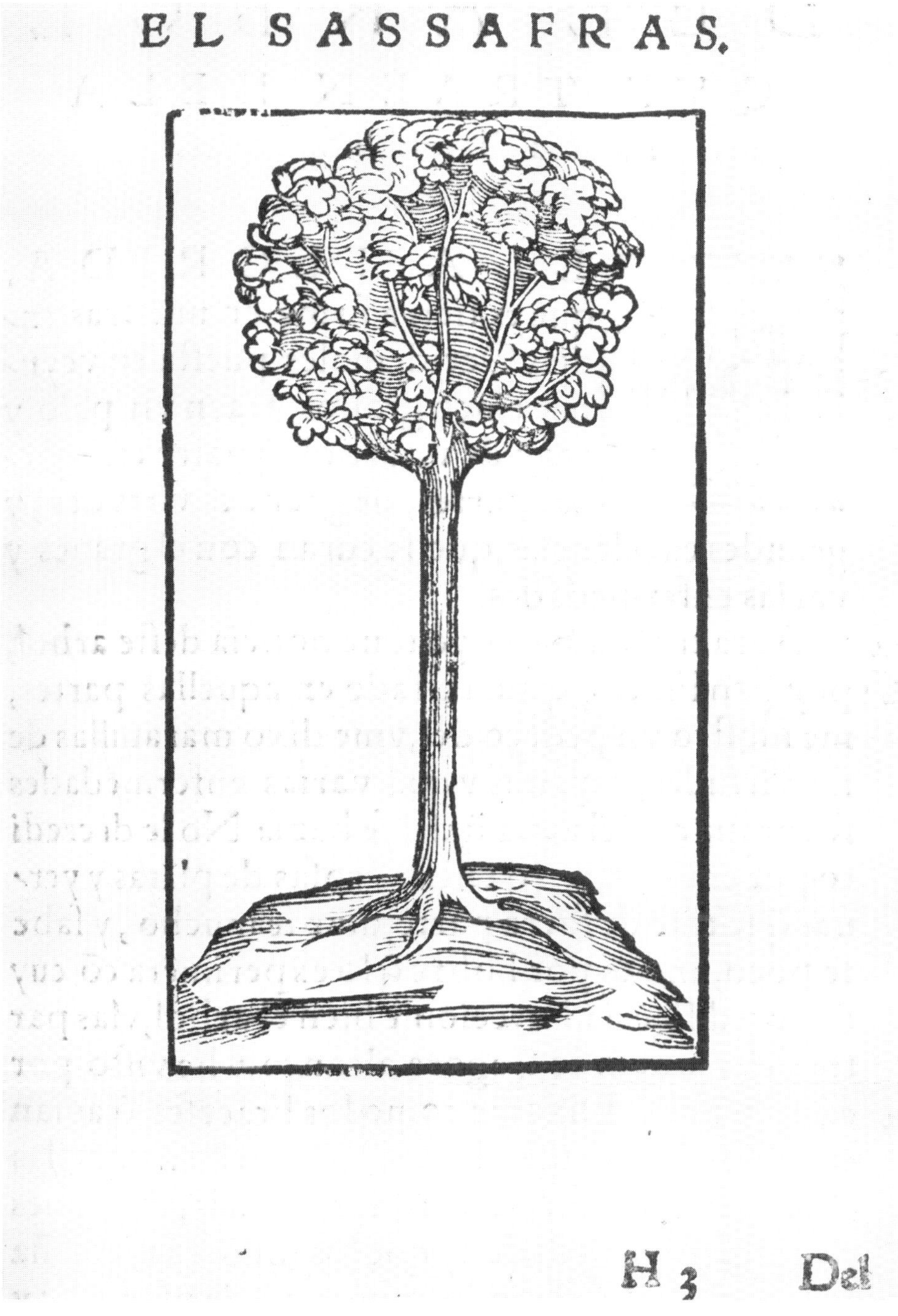

H 3   Del

25. Monardes. *Sassafras tree.*

PIMIENTA LVENGA.

25. Monardes. *Hot pepper.*

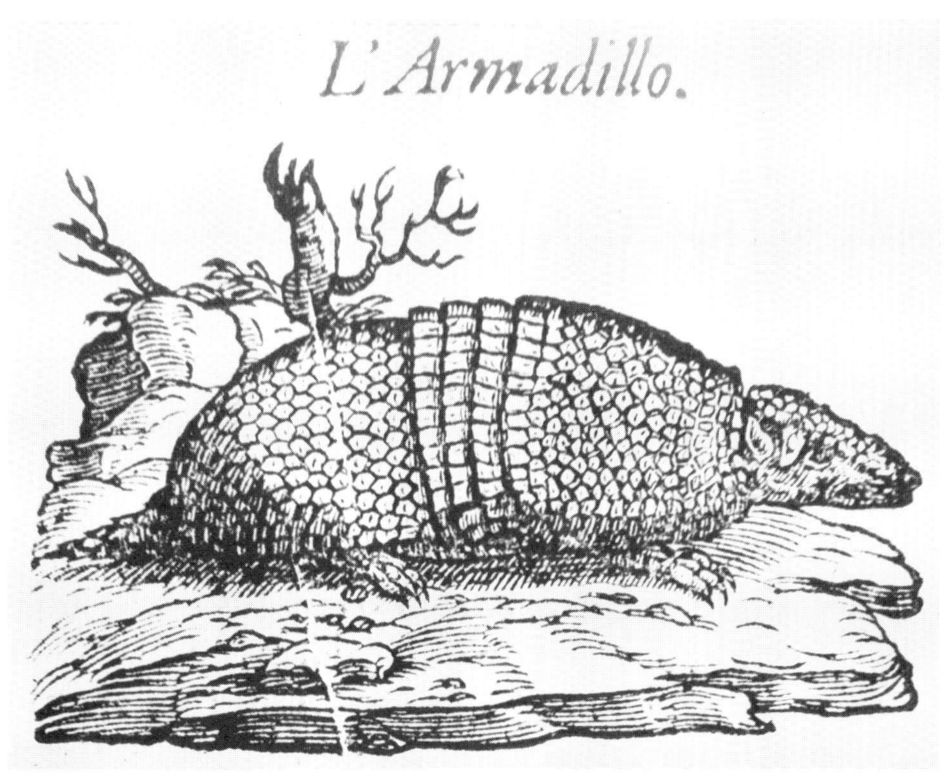

27. Monardes. *Armadillo.*

Del Tabaco, & sue grandi virtù.
Cap.   I.

27. Monardes. *Tobacco plant.*

llamã Canoas. Las quales son de vna pie
ça o vn solo arbol.

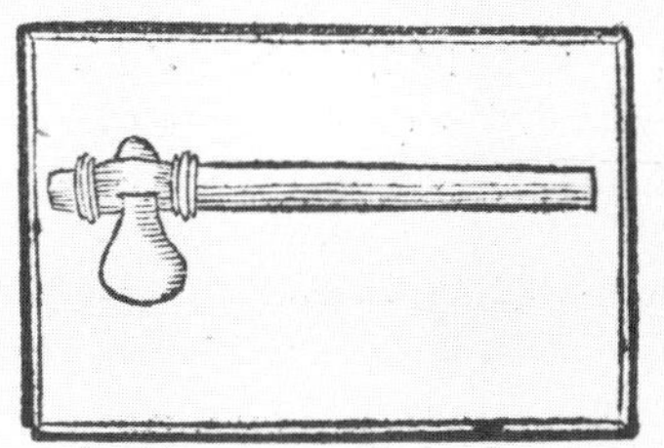

Ablando Plinio en las cosas
de india oriiental dize libro.vj.
capitu.veynte y tres/ que Mo
dusa ciudad esta en vna region
llamada concionada: desde la
qual region se lleua la Pimienta al puer
to llamado Becare: con nauezillas de vn
leño. Estas tales nauetas creo yo que de
uen ser como las que aca vsan los Indios
que son de aquesta manera. En esta Ysla
Española y en las otras partes todas de
estas indias que basta el presente se saben/
en todas las costas dela mar y enlos rios
que los Christianos ban visto basta ago
ra/ay vna manera de varcas q̃ los indios
llaman Canoa/con que ellos nauegan por
los Rios grandes / z assi mismo por estas
mares de aca: delas quales vsan para sus
guerras y saltos:para sus contrataciones
de vna Ysla a otra/o para sus pesquerias
y lo q̃ les conuiene. E assi mesmo los chri
stianos que por aca biuimos no podemos
seruir nos delas credades que estan enlas
costas de la mar y de los rios grandes sin
estas Canoas. Cada Canoa es de vna so
la pieça o solo vn arbol. El qual los indios
vazian con golpes de bachas de piedras
enastadas / como aqui se vee la figura de
llas y con estas cortan o muelen el palo ao
candolo:y van quemando lo que esta gol

peado y cortado poco a poco: z matando
el fuego torñã a cortar y golpear como pri
mero/y continuando lo assi bazen vna bar
ca / quasi de talle de Artesa o dornajo: pe
ro bonda z luenga y estrecha: tan grande
y gruessa como lo sufre la longitud z latti
tud del arbol de que la bazen: y por deba
xo es llana/y no le derã quilla como a nue
stras varcas y nauios. Estas be visto de
porte de quarenta y cinquenta hombres:
y tan anchas que podria estar de traues
vna pipa bolgadamente:entre los indios
Caribes flecheros: Porque estos las v
san tan grandes o mayores como be di
cho z llaman las los Caribes Piraguas:
y nauegan con velas de algodon/y al remo
assi mismo con sus Nabes (que assi llamã
alos remos) Y van algunas vezes bogan
do de pies y a vezes sentados:z quãdo qui
eren de rodillas. Son estos Nabes como
palas luẽgas: y las cabeças como vna mu
leta de vn coxo:segũ aqui está pintados los
nabes oremos:y canoa. Ay algũas destas
Canoas tã pequeñas que no cabẽ sino dos

29. Oviedo. *Indian mattock and canoe, or pirogue.*

29. Oviedo. *Indians panning for gold.*

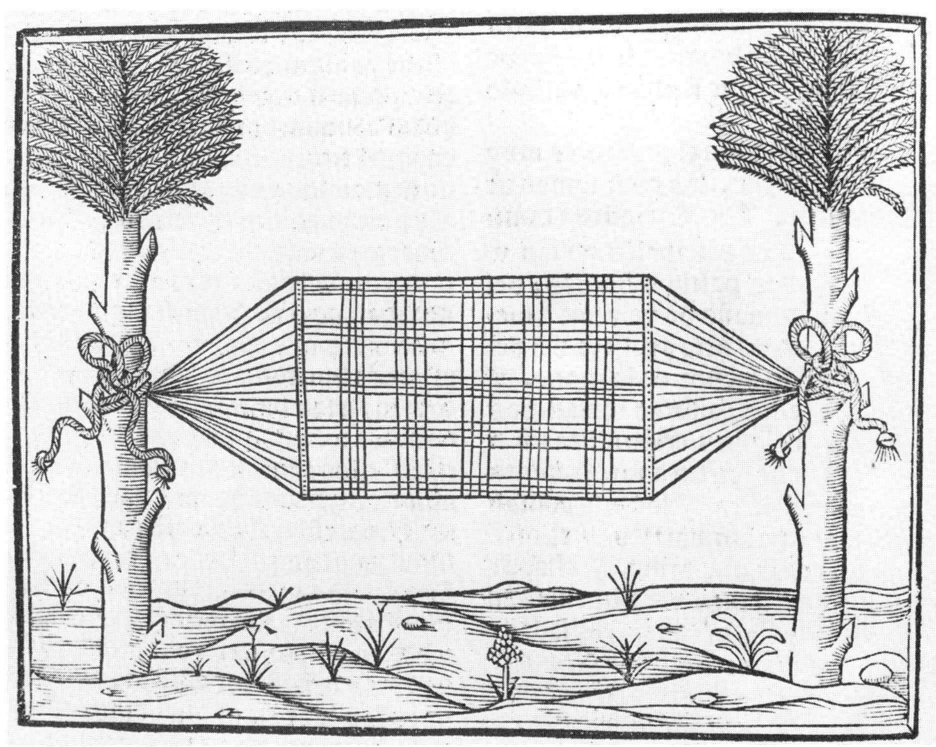

29. Oviedo. *Hammock.*

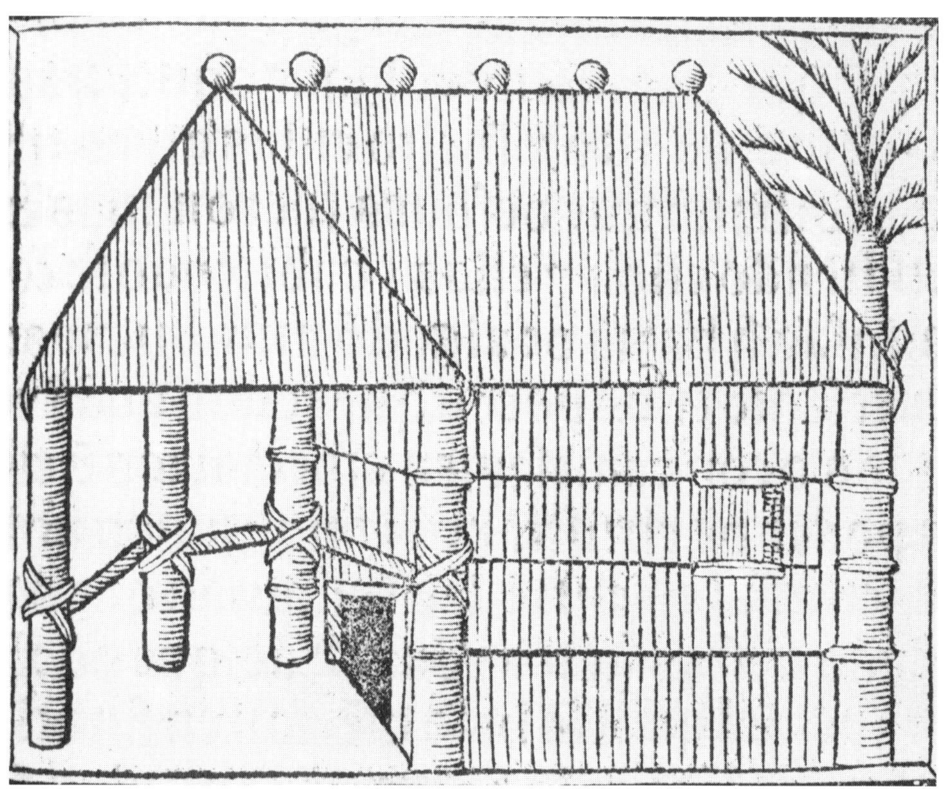

29. Oviedo. *Indian house construction.*

29. Oviedo. *Pineapple.*

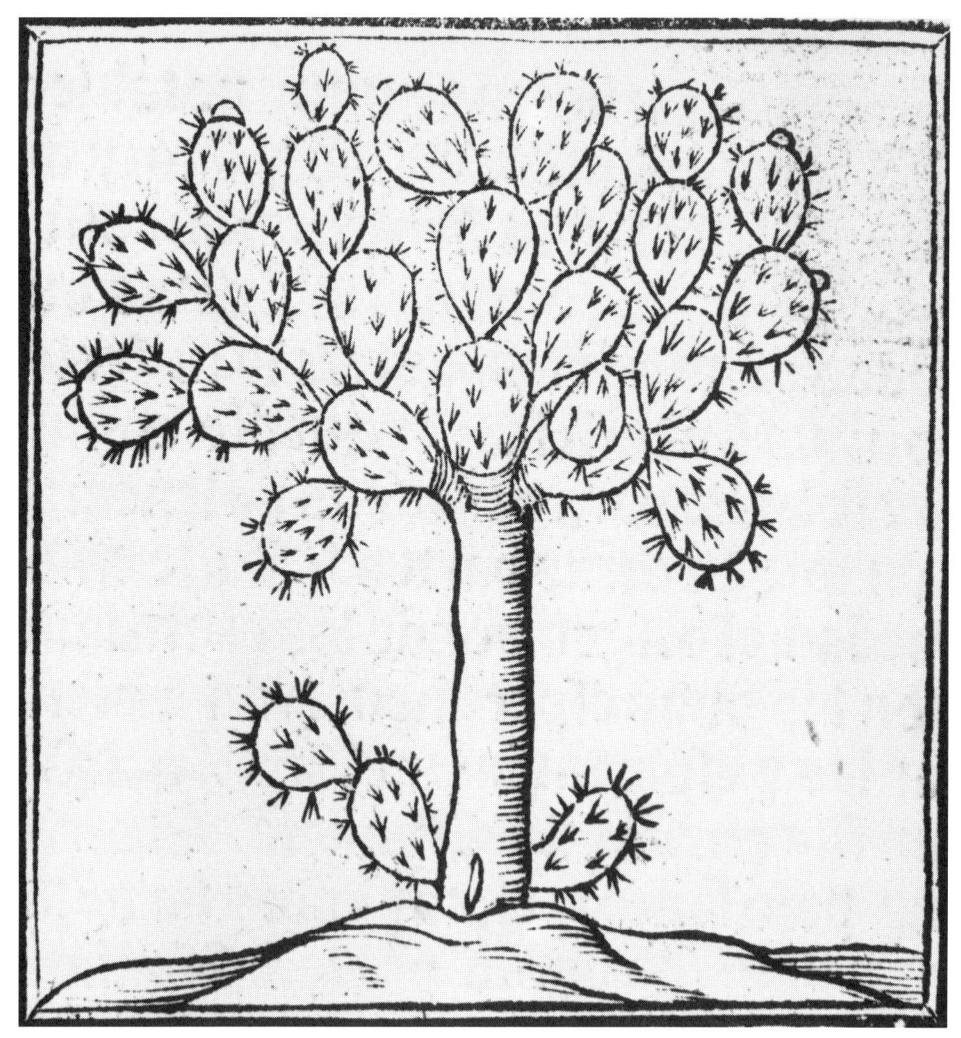

29. Oviedo. *Prickly pear.*

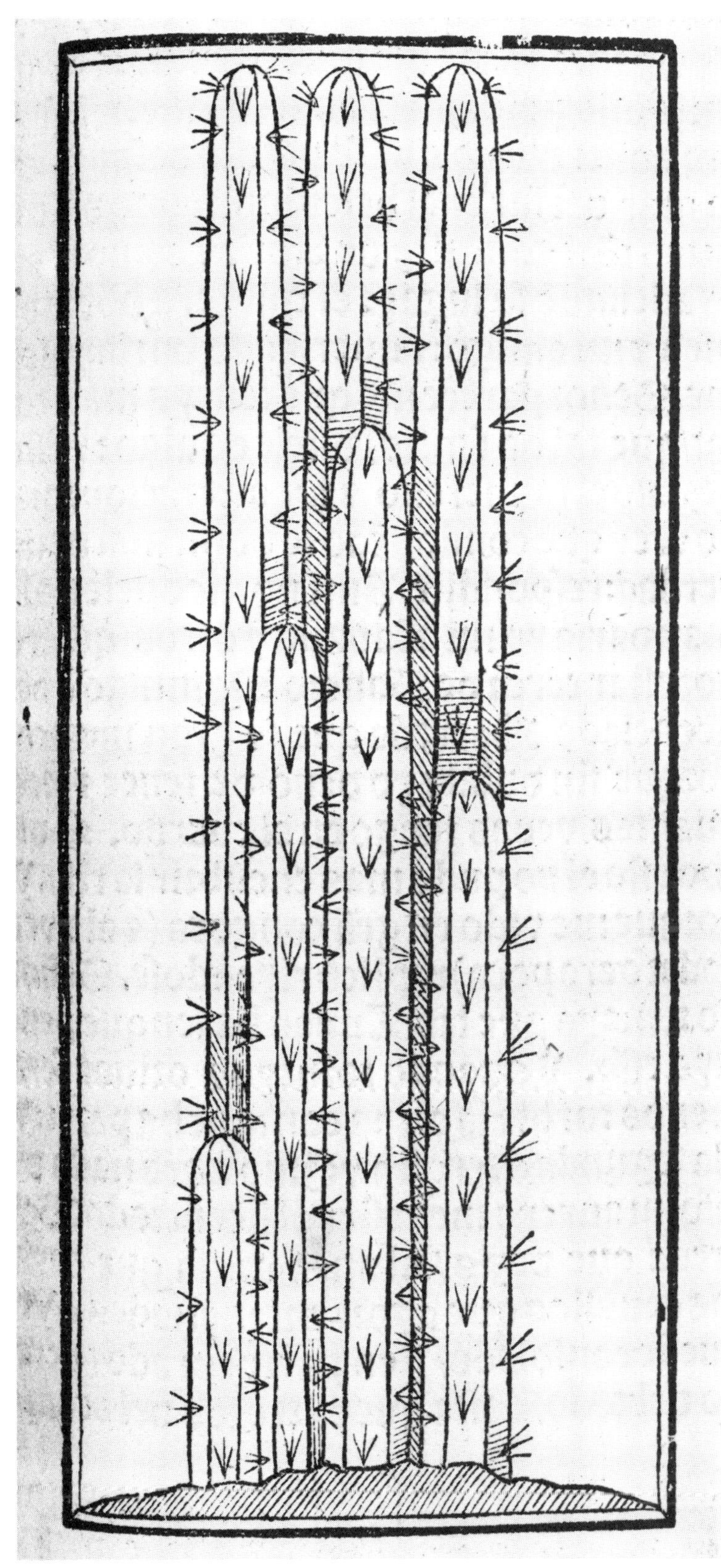

29. Oviedo. *Cactus.*

33. Cieza de León. *Llamas, the only domesticated large mammals indigenous to the Americas.*

33. Cieza de León. *Indian rites of human sacrifice.*

34. Staden. *Indians fishing*.

Sie haben eyne art holtzes/die heysset Vrakueiba/des trí
cknen sie/vnd nemen sein dan zwey stecklin eyns fingers dick/
reiben eyns auff dem andern/das gibt dann staub von sich/
vnd die hitze von dem reiben stecket den staub an/ Darmit
machen sie fewr/wie diese figur anzeygt.

Warin

34. Staden. *An Indian starting a fire without flint and steel.*

34. Staden. *Indians dance around their European captive.*

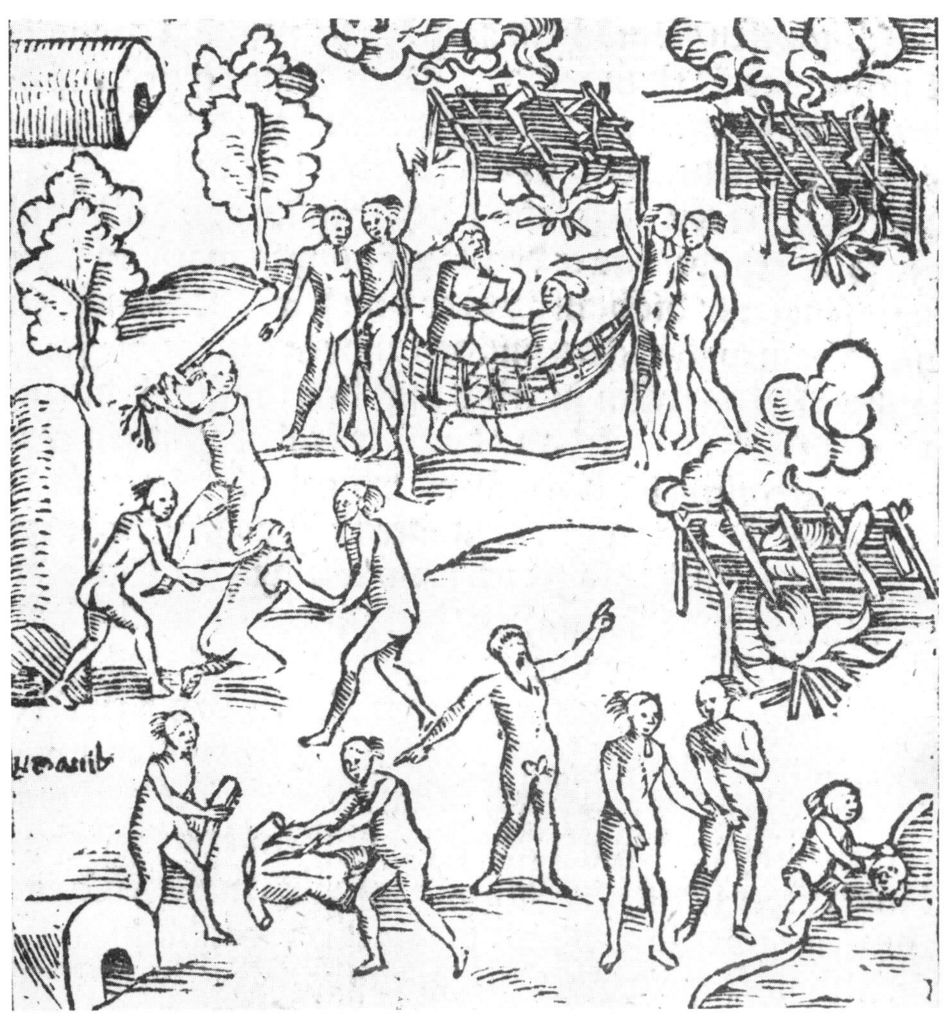

34. Staden. *Indians killing and cooking their captives.*

Vocabulario de la lengua general de los Indios del Peru, llamada Quichua.

A

A Artículo del acusatíuo——————ta, o man
A aduerbío para llamar——— xáy, o xé
A a del q̃ alla a otro en maleficío—— a a
(O athae
A a a ínterjection del que se ríe——— ha, ha, ha,

A. ante. B.

A Bahar, o retener el baho —— áma çama
(ni una,
Abalançarse alguno—— pâuani, pa[illegible]apuc,
Abarca, suela de cuero————— oxòta, [illegible]i. gui.
o pucru
Abarcado calçado cõella— oxotayóc, lla[illegible]isca,
Abarca, calçado de madera——— cillu, o cro,
Abarcar, casí abraçar ——————— márcani. gui
Abarcar como tierra——————— mícllani. gu[illegible]
Abarraganado varon con soltera — sí pas yóc,
(o sí pas yaóc
Abarraganada hembra con soltero— guáyna
yôc, o guaynallicoc,
Abarraganada de casado——————— guáchoc
Abarrar ccmo ala pared ———— chocani. gui,
Abarrancarse el ganado —— tósconi. gui. o
(tóx coni. gui,
Abarrancado el ganado ———— tóx cosca,
Abastar ———— ——— cáma payane
Abasto, o abastan —————— cáma payasca

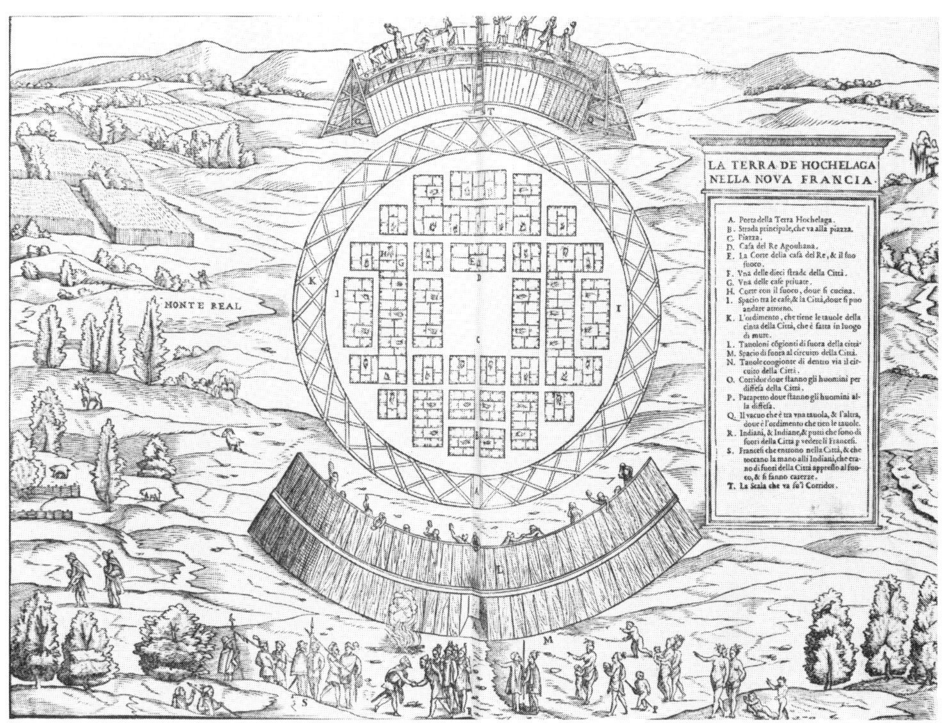

37. Ramusio. *Indians near Montreal defending their village against the French.*

Discovering the Earth's Surface

Maps and Navigation

Europeans from the most ancient times had known that the world was a sphere, but none had ever imagined that a continent separated Asia from Europe. The Columbian voyages, especially the third voyage that explored the coast of Venezuela in 1498, revolutionized European concepts of geography and provided new data for the science of navigation.

As they acquired new information by exploring the Americas, pilots, map-makers, and scientists over the next century were able to improve estimates of the earth's circumference, determine the true extent of the earth's land masses and oceans, and establish the longitude and latitude of most ports and cities all over the globe.

Changes did not come easily. Instruments were rudimentary and timepieces unreliable. Navigators who achieved splendid practical results by dead reckoning also made and recorded celestial observations that were wildly inaccurate. The Portuguese king insisted that his captains use an erroneous method of determining longitude. Many publishers did not bother to update maps when they issued new editions.

Columbus and other captains depended on experienced pilots who conducted their Spanish voyages to the Americas by dead reckoning. Working from route charts, they took advantage of wind and sea currents in the Atlantic and drew on their years of experience at sea to calculate fairly accurately the north-south distance (latitude) traveled. Longitude was more difficult to estimate. Nevertheless, by the year 1500, dozens of Spanish voyages had traveled safely to and from the Western Hemisphere, piloted by men like Juan Aguado, Columbus and his brother Bartolomé Colón, Juan de La Cosa, Alonso de Hojeda, Andrés de Morales, Peralonso Niño, Vicente Yáñez Pinzón, Bartolomé Roldán, Alonso de Torres, and Juan de Umbria. These were just some of the better-known pilots and navigators. There were many, many more.

In the beginning of transatlantic navigation, few records were made public. Ship's logs were not kept until the sixteenth century—Columbus's diary from his first voyage being the great exception. A few Portuguese pilots kept records that they turned over to their king for inclusion in the royal cosmographical charts and tables. But these charts were state secrets not shared with other nations. As a result, we know very little of how any one of these voyages was navigated.

Despite this policy of secrecy, the published works displayed here indicate great changes in methods of navigation. The first changes made celestial navigation practical for seafaring. For centuries, land travelers in desert regions had observed the stars. Like seafarers, desert travelers had no landmarks by which to orient themselves, and a sophisticated science of celestial navigation had been developed by Arab scientists. Celestial observations were extremely difficult at sea, however, and the Atlantic offered no handy islands along the way to take steady, accurate readings. To overcome these difficulties, Spanish and Portuguese pilots and mathematicians invented new instruments for observations, improved old ones, added new data from the Southern and Western Hemispheres, and compiled tables of altitude, longitude, and latitude. Increasingly, seamen made use of celestial navigation in transatlantic voyages, although dead reckoning remained essential until the modern development of accurate and reliable timepieces.

In order to define their position on the earth's surface in circular measure, seamen needed a common set of coordinates for longitude and latitude. The method for measuring latitude from the altitude of the sun or a circumpolar star at meridian altitude was developed during the fifteenth century.

Longitude, however, proved much more difficult to calculate. Since ancient times, scientists understood that the earth's rotation is synonymous with time. Longitude could be measured by timing an astronomical event, such as an eclipse or the conjunction of planets, that could be seen simultaneously by observers in different places. By extrapolating from the time difference, they could estimate the distance separating them. Arab mathematicians in Spain, in fact, had drawn up tables showing the positions of a heavenly body at regular intervals in time when observed from Toledo, so that they could be used in other places.

The problem was that the timepieces of the day were not reliable enough to calculate time accurately. The rate of the earth's rotation is such that one degree of longitude corresponds to four minutes of time. By the year 1500 the best mechanical clocks were subject to an error of about ten minutes a day. Columbus tried to determine longitude by observation of eclipses while at anchor in the Caribbean in 1494 and 1504 using published tables. He miscalculated the true longitude in 1494 by 22°30′; in 1504 by 38′.

With errors of this magnitude common, celestial navigation was not practical. The tables and instruments were not adequate for navigation or cartography. In *Portuguese and Spanish Attempts to Measure Longitude in the 16th Century*, W. G. L. Randles notes that the Portuguese king ordered pilots to use an erroneous master chart of the route to India on which longitudes had been determined by lunar and solar eclipses. The result had been shipwrecks. Portuguese pilots therefore had their charts made secretly in Spain, using traditional methods.

39.
Laurentius Corvinus, 1465?–1527. *Cosmographia dans manuductionem in tabulas Ptholomaei . . .* Basel: Nikolaus Kessler, 1496.

The *Geography* of Ptolemy, originally written in the second century, was not known during most of the Middle Ages. A Byzantine monk, Maximus Planudes, (ca. 1260–1310) discovered a copy of the *Geography* in 1295. It had no maps, so he reconstructed them from the coordinates in the text. In 1405, the Florentine humanist Jacobus Angelus made the first Latin translation and changed the title to *Cosmography*.

Ptolemy's organization of his work into eight books became the model for all geographies for centuries after the Latin translation first appeared. In Book 1, Ptolemy stated his objectives: to give coordinates of places and geographical features and make recommendations for creating a world map and regional maps. The remainder of the book gives names and coordinates for all the *oikumene* (known world): Books 2 and 3 cover Europe; Book 4, Africa; Books 5 through 8, Asia and a summary. Though the map (if it ever existed) did not survive, the text provided tools essential for drawing one: a system of coordinates based on longitude and latitude, instructions for making rigorous maps, and a critical scientific attitude.

Laurentius Corvinus edited Ptolemy's *Geography,* collating the Greek place names with Latin place names. Like all editors of the *Geography,* he was tantalized by Ptolemy's list of place names from ancient texts whose locations were not known—the *Terrae incognitae.*

40.
Pomponius Mela, fl. 43–50. *Cosmographia pomponii cum figuris.* Salamanca: 1498.

Despite the official policy of keeping navigational information secret, scientific curiosity about the newly explored territory was so intense that scholars managed to piece together reports from ship's officers and integrate the information into new views of the earth.

The cosmographical treatise by the ancient writer Pomponius Mela was a popular textbook for undergraduates at the University of Salamanca in the late fifteenth century. This edition by Francisco Núñez de la Yerva, professor of medicine at the university, placed emphasis on teaching students the longitude and latitude of Europe's large cities and ports. The didactic tone of the edition is so intense that, on the title page, Núñez gives the longitude and latitude of Salamanca, where the book was printed: *cuius loci elongatio ab occidenti. IX & ab equinoctiali xlj. gradibus constat.*

The book is open to the famous first Spanish world map. The map is a graduated planisphere showing longitude and latitude of the major ports and cities.

41.
Gregor Reisch, d. 1525. *Margarita philosophica.* Freiburg: Johann Schott, 1503.

Gregor Reisch published this compendium of information about all parts of the world ten years after news of Columbus's first voyage was published throughout Europe.

The world map included in Reisch's encyclopedia shows many Chinas (Sinarum Regio, Sinarum Gangeticus, Sinus Persicus, Sinus Arabicus, Sinus Hisperioris) and many Indies (India Indus, India Intra Gangem Fluvium o Bragma, India Extra Gangem Fluvium). Reisch labels the areas south of India as "Extra Gangem Fluvium" and north of Scithia [modern Siberia] as lands of cannibals (Anthropophagis).

Reisch's world map does not show Columbus's Indies, but some scholars believe that a legend at the bottom of the map may refer to the New World: "Here is not land but sea, in which are many unknown islands, according to Ptolemy."

42.

Jan Glogowczyk, 1445–1507. *Introductorium compendiosum in Tractatum Spere materialis magistri Joannis de Sacrobusto . . . per magistrum Joannem Glogoulensem . . .* Crakow: 1506.

This is the earliest printing of a small work that was to assume great importance in the history of early nautical astronomy, *De sphaera mundi.* It was written in the early thirteenth century by the Englishman Johannes de Sacrobosco, or John of Holywood, probably as an introduction to a more advanced course on astronomy and cosmology. It became a standard text during the age of discovery.

The treatise is based largely on the work of the great Arab astronomer Alfraganus (Al Farghani, d. 861) and presents the Ptolemaic system as put forward in Alfraganus's *Almagest.* It gives proofs of the sphericity of the earth; it defines astronomical and terrestrial terms such as the ecliptic, equator, small and great circles, meridians, astronomical coordinates, and the twelve signs of the zodiac; and it explains various astronomical phenomena, such as eclipses.

It was Alfraganus whose calculation of the degree of the meridian—56 $^2/_3$ miles— so misled Columbus, who failed to take into account the difference between the Arabic and the Italian mile.

43.

Fracanzano da Montalboddo. *Paesi nouamente retrouati. Et nouo mondo da Alberico Vesputio Florentino intitulato.* Vincenza: H. and G. M. de Sancto Ursio, 1507.

After his short sojourn in Portugal, Amerigo Vespucci returned to Spain, became a Castilian citizen, and was appointed to the new post of Pilot Major. The Pilot Major's job was to maintain and update the master sailing chart for Castile, to certify pilots, and to inspect nautical instruments for accuracy.

Vespucci's appointment to this office is something of a puzzle. Christopher Columbus recommended Vespucci to Queen Isabel as an experienced business agent and ship's chandler. Vespucci, as far as we know, never piloted a ship. Long before he became Pilot Major in August 1508, dozens of Spanish pilots had already navigated successful voyages to and from the Americas. There were plenty of competent pilots and navigators available to the crown of Castile. King Fernando probably had mixed motives for appointing him; Vespucci claimed to have solved the problem of determining longitude, and if the king did not keep him happy in Castile, Vespucci might sell information about Spanish explorations and sailing charts to the Portuguese.

The book is open to an illustration for finding the altitude of Canopus from two different positions. A first-magnitude star in the constellation Carina, Canopus is the second brightest star in the heavens.

44.
Martin Waldseemüller, 1470–1521? *Cosmographiae introductio com quibusdam geometriae ac astronomiae principiis ad eam rem necessariis Insuper quattuor Americi Vespucij nauigationes.* Saint Dié: Gautier Lud, 1507.

With Columbus's voyage, mariners and scholars realized that Ptolemy's text, and consequently the maps that had been drawn from it, contained many gaps and errors. Ptolemy made the earth too small, treated the Indian Ocean as closed, omitted sub-Saharan Africa, and was ignorant of the Americas.

In their introduction to their planned world map, Waldseemüller and his colleagues note the new continent, which they label America because they mistakenly thought Amerigo Vespucci was the first to describe it.

45.
Fracanzano da Montalboddo. *Itinerarium Portugale[n]siu[m].* Milan: Joannes Angeles Scinzeler, 1508.

In light of Portuguese explorations, Fracanzano published a map of Africa in which the continent is accurately shaped and fully rounded, though India and the rest of Asia remain vague and malformed.

46.
Claudius Ptolemaeus, fl. 127–151. *Geographia.* Rome: Bernardino dei Vitali, 1508.

Early in the age of discovery, geographers found a way to correct Ptolemy's errors without altering his text; they included updated tables of longitude and latitude for many places not known to Ptolemy. For example, Columbus's discoveries were added to the updated maps.

This reprinting of the 1507 edition has added a world map by Johan Ruysch and an appendix by Marco of Benevento. The list of place names with their latitudes and longitudes has grown to thirty-six folios. Ruysch labeled American sites with the names Columbus had given them.

47.
Martin Waldseemüller, 1470–1521? *Cosmographie introductio.* Strasbourg: Johann Grüninger, 1509.

The treatise on cosmography, which Martin Waldseemüller and Matthias Ringmann wrote as an introduction to a planned world map, was one of the earliest to incorporate information about the Western Hemisphere. This little book became very popular and went through several editions after it first appeared in Saint-Dié in 1507. Two private citizens of the city of Strasbourg published this new edition in 1509.

The earth is depicted in a cordiform (heart-shaped) image and marked in degrees of latitude and longitude. The two parts of the recently discovered American continents are separated by a strait, with the western coast depicted for the first time but without delineation.

48.
Pietro Martire d'Anghiera, 1435–1526. . . . *Opera legatio babylonica occeani decas poemata epigrammata.* Seville: Jacobus Croumberger, 1511.

The Italian humanist Peter Martyr d'Anghiera made his career as a tutor to the children of Fernando and Isabel, as their ambassador on various occasions, and as their official historian of the Americas. In fact, his real job was to publicize the achievements and claims of the Spanish monarchs in elegant Latin that would be admired and believed throughout Europe. Because he was their public relations agent, the monarchs gave him access to many letters, reports, and maps that they otherwise kept secret.

D'Anghiera's *Eight Decades on the New World* include this Spanish map of the Americas. These maps, compiled and maintained by the Spanish government, were not based on Ptolemy's directions or information. Instead, the Spanish mapmakers used information collected by Spanish pilots. The anonymous maker of this map emphasized the shape of the Caribbean Islands, particularly Cuba, Jamaica, Española, and Puerto Rico (San Juan), and the coasts of North, Central, and South America. The South American river that he labels as Rio Grande is the Orinoco River, the delta of which Columbus had explored and described in 1498.

49.
Claudius Ptolemaeus, fl. 127–151. *Geographia.* Venice: Jacopo Pencio, 1511.

Venice became an early center for the publication of classical Greek works. Venetian printers were among the first to publish Ptolemy's *Geography* in the early fifteenth-century Latin translation by Jacopo d'Angelo. In 1511, Bernardo Silvano of Eboli edited the *Geography* with several innovations; he provided a new title page, included a poem by Giovanni Arelio Augurello, appended a description of the New World by Marco of Benevento, and added a world map by Johan Ruysch.

The book is open to Ruysch's world map, which is in cordiform image. Ruysch labels South America with names derived from Columbus's report on his third voyage, when he speculated that the great rivers of the region might be the four rivers of life near the Terrestrial Paradise. North of this Terra Sanctae Crucis, Ruysch draws a line and labels the territory "Cannibalus."

50.
Claudius Ptolemaeus, fl. 127–151. *Geographia.* Strasbourg: Johann Schott, 1513.

This edition was prepared by Martin Waldseemüller, Matthias Ringmann (Philesius), Essler, and Uebelin. The text includes letters by Pico della Mirandola, Essler, Uebelin, and Giraldi (Ziraldus).

After completing the *Cosmography,* making wood-block gores for a globe, and printing the sea chart and the world map, the humanists at Saint-Dié resumed work on their long-delayed edition of Ptolemy's atlas, to be entitled *The geographical work of Claudius Ptolemy, the man of Alexandria, a most erudite philosopher of mathematical learning, closely reprinted by reproduction of the Greek originals.* In order to resolve disparities in place names and textual content, the scholars turned to other manuscript and printed copies of Ptolemy's atlas. The classical scholar Matthias Ringmann translated place names and texts from Greek to Latin. Waldseemüller designed the maps.

This is the book that gave the name America to the newly discovered lands in the West. It includes Vespucci's letter to Piero Soderini, dated from Lisbon on 4 September 1504, in which he describes all four of the voyages he claimed to have made.

The book is open to one of forty-five double maps that Waldseemüller designed for this edition.

51.
Martin Waldseemüller, 1470–1521? *Cosmographiae introductio cum quibusdam geometriae ac astronomiae principiis ad eam dem necessariis.* Lyons: Jean de La Place, 1517–1518?

During the ten years after their first publication in Saint-Dié, Waldseemüller and his colleagues accumulated much additional information about the Spanish and Portuguese voyages of discovery. They revised some of their earlier misconceptions, realizing, for example, that the lands discovered by Columbus and those reported by Vespucci were the same.

On the South American continent, they deleted the name "America" and inserted the labels "Brazil, land of parrots," and "all of this province was discovered by mandate of the king of Castile." They also deleted the strait between the two American mainlands.

Although only one set of the revised world map survives, the new information was incorporated into the *Cosmography's* navigational information in later editions. This 1517–1518 edition from Lyons is open to a chart for calculating position.

52.
Pomponius Mela, fl. 43–50. *De orbis situ libri tres . . .* Basel: Andreas Cratander, 1522.

Wishful thinking continued to dominate map making outside the Iberian Peninsula. In this edition of Pomponius Mela's geography, the publisher in Basel added a foldout map displaying a fictitious strait through the Isthmus of Panama.

53.

Marco Polo, 1254–1323. *Cosmographia breue introdutoria en el libro de Marco Paulo. El libro d'l famoso Marco Paulo Veneciano d'las cosas marauillosas que vido en las partes orientales. Conuiene saber en las Indias. Armenia. Arabia. Persia & Tartaria. E del poderio del gran Can y otros reyes. Con otro tratado de Micer Pogio Florentino que trata de las mesmas tierras & yslas.* Seville: Juan Varela, 1518.

Just as the ancient geographers Ptolemy and Pomponius Mela continued to command respect, despite their obvious gaps and errors, the medieval merchant explorer Marco Polo enjoyed a resurgence of popularity. Italian humanists such as Paolo dal Pozzo Toscanelli and Poggio Bracciolini attempted to draw geographical conclusions from Polo's descriptions of Asia. This led to misconceptions—and momentous consequences.

Toscanelli, for example, read Marco Polo's description and interpreted it to mean that China extended much farther into the ocean than is the case, and that Japan lay one thousand miles from the Chinese coast. Convinced that most of the globe's surface was covered with land rather than water, Toscanelli had, in 1474, mistakenly informed the Portuguese king that only 8,125 miles separated Lisbon from the east coast of Asia. He also assured the Portuguese king that Japan and other Asian islands were handily close for resupply on a westward voyage.

Christopher Columbus got hold of a copy of this letter, and he believed it. It formed the basis for his early proposals for a westward voyage to Asia, proposals that were rejected by scientific commissions in Portugal and Spain. Even after Columbus explored the coast of Venezuela in 1498 and concluded that it was a heretofore unknown continent, he reinterpreted Marco Polo to turn this obstacle into an incentive. If he could find a strait around this new continent, which he tried to do on his fourth voyage, the distance he would have to travel between its west coast and Asia would be all the shorter.

This hope for a small ocean between South America and China gained credence with every new publication of Marco Polo's work, because his own estimate of the distance between China and Japan was even greater—fifteen hundred miles. Such optimistic estimates were the navigational foundation on which the Magellan expedition was planned, outfitted, and provisioned, with tragic consequences when the fleet had to cross a Pacific Ocean thousands of miles wider than anticipated.

The book is open to the title page, with woodcut illustrations of Marco Polo, Poggio, Santo Domingo, and Calicut—the commercial port on the southwest coast of India that had been the ultimate objective of Columbus and the Portuguese.

54.

Hernando Cortés, 1485–1547. *La preclara narratione di Ferdinando Cortese . . .* Venice: Bernardinus Vercellensis, 1524.

In this Italian edition of Cortés's map of Mexico and Mexico City, the publisher added a useful bit of information in the legend at bottom left. He provided a scale of miles and explained how to convert leagues into miles.

55.

Simon Grynaeus, 1493–1541. *Novus orbis regionum.* Basel: Johann Herwagen, 1532.

Maps and geographies have always depended on the talents and expertise of specialists in many fields. This book was edited by Simon Grynaeus from data collected by Johann Huttich. The emphasis in the text is on Huttich's calculations of longitude and latitude. Unfortunately, they did not take into account the new information brought back by Spanish explorers. Their map of North and South America shows the two continents separated by a strait, and the east coasts of North and South America are crude in comparison with earlier printed maps.

56.

Claudius Ptolemaeus, fl. 127–151. *Geographia.* Basel: Heinrich Petri, 1542.

By 1542, much of the American coastline had been accurately mapped. Here, although the cartographer has rendered the shapes of North and South America more accurately than in earlier editions, there are still many gaps and errors.

57.

Claudius Ptolemaeus, fl. 127–151. *Geographia.* Basel: Heinrich Petri, 1545.

Cartographers updated the maps of individual countries from one edition to another of the *Geographia,* but the original version of Ruysch's world map continued to be in use long after new information had made it obsolete.

58.

Claudius Ptolemaeus, fl. 127-151. *Geographia.* Venice: Nicoló de Bascarini, 1548.

In this Venetian edition of Ptolemy's *Geography,* the publishers have included updated instructions and a chart to show "how it is possible to describe the world in two dimensions as accurately as if it were drawn on a sphere." Ptolemy's instructions are for a cylindrical projection, which could not preserve the true angular relationships between places. Navigators who tried to steer a course measured on this type of projection were forced to make constant corrections in order to be assured of a reasonably accurate landfall.

59.
Johannes Honter, 1498–1549. *Rudimentorum cosmographicorum.* Zurich: Christoph Froschauer, 1549.

For the convenience of travelers, Johannes Honter published his *Rudiments of Cosmography* in pocketbook size. His world map is in cordiform projection to accommodate longitude and latitude, and, still using outdated information, he shows a strait through Panama.

60.
Henricus Glareanus, 1488–1563. *De geographia.* Venice: Pietro, Giovanni, Maria, and Cornelio dei Nicolini da Sabbio, 1549.

This geography lagged even farther behind in its knowledge of place names and coordinates. Although the author entitled his final chapter "On the regions not in Ptolemy," he describes America as consisting of two islands, "Spanolla" and "Isabella."

61.
Sebastian Münster, 1439-1552. *Cosmographiae universalis.* Basel: Heinrich Petri, 1550.

The Basel scholar Sebastian Münster's *Universal Cosmography* assembled maps, images of cities, and descriptions of lands and peoples of the world, including the Americas as well as Asia and Europe. The book was a best seller, although Münster did not discard the old to accommodate the new.

The book is open to his "Map of the New Islands." South America is labeled "The New World," yet it is treated as part of Asia.

62.
Francisco López de Gómara, 1510-1560? *. . . Historia general de las Indias con todo el descubrimiento y cosas notables que han acaecido dende que se ganaron ata el año de 1551. Con la co[n]quista de Mexico y de la nueua España.* Zaragoza: Agustín Millán, 1553.

Spaniards never forgot that their original intention had been to find a westward route to Asia. This 1553 edition of Gómara's history of Mexico includes up-to-date maps of North and South America, as well as a map showing the principal stops along the commercial route from Europe, around Africa, to Asia: San Lorenzo, the Red Sea, Calicut, Ceylon, Malacca, Sumatra, Borneo, the Moluccas, and China.

63.
Pedro de Medina, 1493–1576. *Arte de navegar en que se contienen todas las reglas, declaraciones, secretos, y avisos, que a la buenanauegacion son necessarios, y se deuen saber, hecha por el maestro Pedro de Medina.* Valladolid: Francisco Fernández de Córdova, 1545.

Pedro de Medina was a teacher of mathematics and a founder of marine science. In 1538, he received permission from Emperor Charles V to draw charts and prepare pilot books and other devices necessary for navigation to the Indies. He made charts, sailing directions, astrolabes, quadrants, mariner's compasses, and cross-staffs.

The Pilot Major of Castile, Sebastian Cabot, accused Medina of constructing and selling unauthorized charts. Medina, confident of his scientific expertise, retaliated. In the judicial hearing he criticized the scientific validity, accuracy, and precision of the charts, astronomical tables, and instruments approved by Cabot. He convinced the judges of Cabot's technical incompetence and scientific ignorance, and they eventually ruled in Medina's favor. In 1549, the monarchy appointed Medina "honorary cosmographer."

In 1545, Medina published his *Art of Navigating,* of seminal importance to the development of seafaring. Though its principles can be traced to unpublished Portuguese and Castilian manuals, this was the first time the Castilian navigational secrets were published in full.

64.
Martín Cortés. *Breue compendio de la sphera y de la arte de navegar con nueuos instrumentos y reglas exemplificado con muy subtiles demonstraciones.* Seville: Anton Alvarez, 1551.

Martín Cortés (no relation to the conqueror of Mexico) was born in the town of Burjalaroz in the Kingdom of Aragon. He moved to Castile and became a citizen of the city of Cádiz, an important center of cosmographic research.

Cortés drew on Spanish and Portuguese naval experiments to propose better methods for charting position by observed latitudes. One of his important contributions to navigation was to suggest formulae for correcting bearings for magnetic variation.

65.
Gemma Frisius, 1508–1555. *De principiis astronomiae.* Antwerp: Joannes Graphaeus, 1553.

Gemma Frisius, Dutch mathematician, made several vital contributions to cosmography. His first was to solve, in principle, the persisting problem of how to determine longitude, a serious problem in all oceanic voyaging. In a 1530 publication he proposed a method that eventually became the basis for modern practice. His solution was to carry a timepiece set to local sun time at the point of origin. Comparing the difference between the time shown on the clock and the sun time at the remote location would yield the longitude. Although theoretically correct, Gemma's procedure often resulted in errors when used at sea because the timepieces of the day were not accurate.

The book is open to a chart in which Gemma indicated his own calculations of longitude and latitude for the major cities and places of his day.

66.

Gemma Frisius, 1508–1555. *Les principes d'astronomie.* Paris: Guillaume Cavellat, 1556.

Gemma's second significant contribution to cosmography was a description of the use of triangulation in surveying. The geometrical bases of triangulation were well known, but he was the first to print a full description of the technique for mapping purposes, including the measurement of base line, the means for setting the scale of a map, and the use of resectioning for orienting the instrument—the predecessor of the modern surveyor's compass. Wide dissemination of this treatise launched a new era in the production of large-scale regional maps.

This French edition of Gemma's treatise is distinguished by the fine draftsmanship of the scientific illustrations. The book is open to a demonstration of triangulation in surveying.

67.

Johann Schoner, 1477–1547. *Opera mathematica Ioanni Schoner; Carolostadii in unum volumen congesta, et publicae utilitati studiosorum omnium.* Nuremberg: Johann von Berg and Ulrich Neuber, 1561.

Johannes Schoner was a geographer from Karlstadt who acquired a copy of Waldseemüller's world map and his *Carta marina.* Schoner bound the twelve sheets of the world map together with the twelve sheets of the *Carta marina,* a star chart by Albrecht Dürer, and two incomplete sets of gores, one for a terrestrial globe and another for a celestial globe. Schoner's is the only surviving copy of the world map.

This mathematical and geographical treatise by Schoner is open to an item increasingly popular in published geographical works—a volvelle, a paper contrivance of movable parts for ascertaining the time of the rising and the setting of the moon and sun and the time of high and low tide. Each part of the volvelle was printed on a separate page, so that the reader could cut out the parts or trace them on separate pieces of paper, and assemble the various parts with string.

68.

Pedro de Medina, 1493–1576. *Regimiento de navegacion.* Seville: Simón Carpintero, 1563.

Medina's treatise on navigation enjoyed several printings, partly because of the fine illustrations showing how to use some of the new navigational instruments. Instruction in the use of navigational instruments was not part of his job as honorary cosmographer, but his position obviously made his book desirable to those who hoped to pass the exams given by the Pilot Major.

The book is open to an illustration of how to use the cross-staff, or *ballestilla.* The cross-staff was an instrument used to determine the altitude of the polestar

(Polaris or North Star), when the Guard Stars were in a particular position. The Guard Stars (or Pointers) are the two end stars in the bowl of the Big Dipper.

An observer held the end of the staff to his eye and moved the transversary (crosspiece) along the staff until the upper edge, viewed from the end of the staff, appeared to cut through the celestial body being observed, and the lower edge appeared to lie on the horizon. A scale engraved on the staff enabled the angle (or a function of it) to be read from the position of the transversary.

Several problems confronted an observer trying to use the cross-staff. Medina recognized and discussed the problems of the position of the eye in relation to the axis of the staff (parallax error) and consequential inaccuracies in the observations. Because a cross-staff required the user to look directly at the observed object, it had to be modified for solar observations.

69.
Abraham Ortelius, 1527–1598. *Theatrum orbis terrarum.* Antwerp: Gillis Coppens van Diest, 1570.

Abraham Ortelius was an antiquities merchant in Antwerp, a center of fine printing and advanced map making. He and his sisters took up the coloring of maps and he was admitted to the painters' guild as an illuminator of maps in 1547. Gradually his shop began to specialize in the book and map trade. By 1564, he began publishing maps himself, including an ambitious wall map of the world.

But his greatest contribution to cartography was this book, *Theater of the World*, containing a collection of maps. Ortelius wanted to supply merchant traders in the Low Countries with geographical information so they could calculate freight costs and plan efficient trade routes. For this collection he wanted maps based on data brought back by explorers to the Americas, Asia, and Africa. He was willing to dispense with Ptolemy to achieve his goal.

The *Theater of the World* brings together the newest and best maps available for all parts of the world. Ortelius had them engraved to uniform size and published them together in book form. The book was an instant and continuing success, largely because of its practical nature. His contemporary, Gerhard Mercator, thanked Ortelius for having "selected the best descriptions of each region and collected them into one manual, which can be bought at small cost, kept in a small space and even carried about wherever we please." Because of this format, the *Theater of the World* is considered the first atlas.

The 1570 edition contained fifty-three map sheets. By 1612, the book was being published with 167. Ortelius did not pretend that he had made these maps himself; he scrupulously credited individual cartographers and included a catalogue of all the maps known to him. Because he was a leading collector of antique and modern maps, this catalogue is of primary importance for the history of cartography.

70.

Giasone de Nores, ca. 1530–1590. *Breve trattato del mondo, et delle sue parti, semplici, et miste . . .* Venice: 1571.

Altura navigation developed before latitude navigation, when the Portuguese were making their north–south voyages from Portugal down the west coast of Africa. The return voyage involved a wide sweep westward into the Atlantic to cross the northeast trade winds and regain the variable and westerly winds in the latitude of the Azores. Seamen needed some method of checking on a daily basis the north–south component of the position obtained by dead reckoning, which in turn depended on accurate timekeeping. Altura navigation, estimating time by observing the rotation of the Guard Stars around the North Star (Polaris or Pole Star), developed. A line extended through the Guard Stars leads to the North Star, and the altitude of this line indicates the time of night.

Giasone de Nores advised seamen to observe the North Star on departure when the position of the Guards indicated that it was on the meridian. Portuguese mathematicians gave the height at Lisbon for all eight positions, from which the heights observed later could be subtracted to get the distance traveled north or south. Later tables gave the meridian solar altitude for each day of the year at Lisbon, Madeira, and other important seaports. For example, subtracting the altitude observed from the Lisbon figure would give the distance from Lisbon.

71.

Juan Pérez de Moya, 1513?–1596? *Tratado de cosas de astronomia, y cosmographia, y philosophia natural.* Alcalá: Juan Gracián, 1573.

Juan Pérez de Moya was the most noteworthy mathematician of sixteenth-century Spain. He was born in Santisteban del Puerto in the Sierra Morena. He studied at the universities of Alcalá and Salamanca, and in the later half of the century he became a canon of the cathedral at Granada.

This work is divided into three books. The first treats of the composition of the globe, longitude and latitude, stars, planets, orbits of the sun and moon, and eclipses. The second deals with elements in meteorology, volcanoes, zones, climates, and problems of cartography. Here Pérez de Moya describes America and New Spain, summarizing the history of exploration and Spanish conquest there to date, and describing the Indians and natural history of the continent. In the third book, he treats sundials and other methods of keeping time.

The book is open to a small circular woodcut map (diameter 49 mm) illustrating a problem in navigation.

72.

Pedro de Medina, 1493–1576. *L'art de naviguer de M. Pierre de Medine espagnol contenant toutes les reigles, secrets, & enseignemens necessaires à la bonne nauigation.* Lyon: Guillaume Rouillé, 1576.

Medina's treatise on navigation continued to enjoy great popularity through the sixteenth century. This French edition included an updated sailing chart.

73.

Andrés de Poza, d. 1595. *Hydrografia. La mas curiosa que hasta aqui ha salido a luz, en que de mas de un derrotero general, se enseña la nauegacion por altura y derrota, y la del este oeste: con la graduacion de los puertos, y la nauegacion al Catayo por cinco vias diferentes.* Bilbao: Matias Marés, 1585.

Andrés de Poza was a university graduate in law. Born in the city of Orduña, he practiced law in the Basque Country. By the time he wrote this book, knowledge of the navigable waters of the world had grown enormously, and experts had developed specialties. In this book Poza addressed the specialty of establishing the position and contours of coasts.

The speed and accuracy with which the outlines of the new world emerged on the charts of the time suggests that many Castilian pilots were already familiar with the new science of celestial navigation. Poza incorporated almost a century of mariners' observations at sea in his treatise, which he divided into two books; the first on navigation and astronomy, and the second giving information on the ports, coastlines, bays, and tides of the Atlantic Coast. He also included instructions for five different routes to China.

74.

Washington Irving, 1783–1859. *History of the Life and Voyages of Christopher Columbus.* New York: G. and C. Carvill, 1828. 3 vols.

We have seen that during the Middle Ages people knew the earth was round. But why do so many people today think that medieval Europeans believed the earth was flat?

The answer lies in the modern age and, specifically, in this book by Washington Irving. Irving was already a famous American author when he published this multivolume biography of Columbus. He was also well informed; during the 1820s he spent several years in Spain, with full access to unedited manuscripts and the great Columbus research then being carried out by Spanish maritime historians.

But Irving, dissatisfied with history, wanted a good story and did not mind making it up. He invented a mythical Columbus—a scientific hero who overcame superstitious and ignorant opponents.

Irving knew that the Spanish scientists who rejected Columbus's plan shared the medieval knowledge of the earth's sphericity and estimated the circumference of the earth much more accurately than Columbus had. But that would not have made nearly as good a story.

Ignoring historical evidence, Irving wrote a lively fable of courage versus cowardice, learning versus ignorance, common sense versus superstition. That fable has distorted the history of the Columbus voyages and of scientific discovery and exploration ever since. While scientists and explorers rethought the world in the wake of Columbus's voyages, an American novelist recreated Columbus—and the Middle Ages.

75.

Lodestone. Maker unknown, early 18th century, English magnetite, bound in brass.

Before the discovery of electromagnetism, the only way to make an iron bar magnetic was to rub it with a lodestone, a naturally occurring mineral called magnetite. Chemically this mineral is an oxide of iron with powerful magnetic properties. Who first discovered that a bar of iron rubbed with a lodestone would tend to point toward the north—and so invented the compass—remains a subject of controversy. The chances are that it was someone living in China in the second or third millennium B.C.

The most powerful stones came from China and Bengal. They were bluish in color and were commonly sold for their weight in silver.

The example seen here is small enough to be a practical tool aboard ship. To remagnetize his compass needles, the navigator simply rubbed them against the stone.

From the collection of Raymond J. and Laura Wielgus.

76.

Nocturnal. Maker unknown, English, early 18th century, boxwood and brass.

Nocturnals are devices for telling the time at night. They are based on the fact that the stars, while remaining fixed relative to one another, appear to rotate around the North Star. It is the Earth that is rotating, of course; the North Star appears fixed because it lies along the Earth's axis of rotation. All the other stars appear to rotate, and their position at any moment indicates the time.

This nocturnal is made of several pieces of wood pivoted together at the center so they can rotate relative to one another. The navigator holds the nocturnal upright by the handle until the North Star can be sighted through the hole at the axis of rotation. The navigator then turns the long arm until it lies along the line made by the two brightest stars in the constellation of the Great Bear (also known as the Big Dipper). These two stars, called the Guard Stars, are easily seen and they lie along a line that passes close to the North Star. As an alternative, the bright star, called the Pole Star, of the Little Bear (Little Dipper), can also be used.

This nocturnal can be used with either the Great or the Little Bear, as the words "Both Bears" on the handle indicate.

From the collection of Raymond J. and Laura Wielgus.

77.

Michiel Coignet, 1549–1623. *Instruction nouvelle des poincts plus excellents & necessaires, touchant l'art de naviguer.* Antwerp: Henry Hendrix, 1581.

Michiel Coignet was one of the outstanding instrument makers of the sixteenth century and his *Instruction nouvelle* is one of the important books in the history of navigation. In it Coignet illustrates the navigator's nocturnal for the first time and provides the first description and illustration of a cross-staff having more than one transversary. The primary purpose of the book was to serve as an instruction manual and advertisement for the author's inventions.

From the collection of Raymond J. and Laura Wielgus.

78.
Gunner's rule. Maker unknown, English, before 1635.

This pocket rule helped ships' gunners to calculate the powder and shot needed to fire at various distances and spreads, such as point blank and random.

From the collection of Raymond J. and Laura Wielgus.

79.
Gunter quadrant. Made by John Prujean, fl. 1667, d. 1701. New College Lane, Oxford, brass.

A quadrant was one-quarter (hence the word quadrant) of a full sun dial—an angle-measuring device with a plumb bob and a scale of degrees—used to tell time. This example, made for the latitude of Oxford and constructed in accordance with principles laid down by Edmund Gunter in 1623, offers the navigator two sighting pinnules, a scale from 0 to 90 degrees, and a plumb line for taking the altitude of stars. Quadrants could also be astronomic instruments, with a stereographic projection of the sky (Equator, ecliptic, Tropic of Cancer, and the positions of six stars). On the other side, this example has a shadow square and a zodiacal calendar.

From the collection of Raymond J. and Laura Wielgus.

80.
Edmund Gunter, 1581–1626. *The Description and use of the sector; the cross-staffe and other instruments.* London: William Jones, 1624.

Gunter's contributions to science were essentially of a practical nature. He had a gift for devising instruments which simplified calculations in astronomy, navigation, and surveying, and which put the theory of navigation into a form suitable for use at sea. His extensive treatise illustrates a variety of mathematical principles and instruments. Here it is opened to the woodcut depicting Gunter's quadrant.

From the collection of Raymond J. and Laura Wielgus.

81.
Gunter navigation scale. Maker unknown, English, late 17th century, boxwood.

This scale for solving mathematical problems was exactly two feet long. On one side, the scale was inscribed with eight sets of information: the line of numbers, marked Numbers; the line of artificial sines, marked Sines; the line of artificial tangents, marked Tangents; the line of artificial versed sines, marked V. S.; the artificial sines of the rhumbs, marked S. R.; the artificial tangents of the rhumbs, marked T. R.; the meridian line on Mercator's chart, marked Merid.; and equal parts, marked E. P.

The lines of artificial sines, tangents, and numbers are calibrated on this scale so that the navigator could solve any trigonometry problem in navigation or astronomy with a pair of compasses.

From the collection of Raymond J. and Laura Wielgus.

82.

Samuel Sturmy, 1633-1669. *The Mariners Magazine, stor'd with these mathematical arts: the rudiments of navigation and geometry, the making and use of divers mathematical instruments . . . The third edition, diligently revised and carefully corrected by John Colson.* London: John Playfair, 1684.

Sturmy's *Mariners Magazine* originally was published in 1669, the year Sturmy died. The second edition appeared ten years later, and this, the third and final edition, in 1684. It is open to the plate showing the navigation scale.

From the collection of Raymond J. and Laura Wielgus.

83.

Equinoctial ring dial. Made by Nicolas Bion, ca. 1700, French, brass.

Nicolas Bion (165?–1733) was a successful maker of scientific instruments. His *Treatise on the construction and principal uses of mathematical instruments,* first published in 1709, demonstrates his gift for explaining clearly to the general reader how things worked.

With the equinoctial ring dial, the navigator could establish daylight time at any latitude. The broken lines represent all the possible paths of light at noon, depending on the month. The ring is set to a latitude of 48 degrees, that of Paris.

From the collection of Raymond J. and Laura Wielgus.

84.

Sextant. Richard Lekeux, 137 High Street, Wapping, near Execution Dock, London, late 18th century, ebony, ivory, brass, glass.

The four diamond-shaped ivory inlays are later mariners' work. Mahogany case with later New York [state?] label that was added by the recipient when acquired from the shop.

The sextant was introduced in 1757 to measure lunar distances as part of the continuing problem of establishing longitude. A sextant will read very large angles up to 120 degrees; however, double reflection causes the arc of the instrument to be only half that—60 degrees, or a sixth of a circle, which accounts for its name. The navigator used the sextant to measure lunar distances. The moon moves through an angle of thirteen degrees every twenty-four hours. At any given instant, therefore, the moon is at a certain fixed distance from any star, including the sun. These distances were tabulated in the late 1750s by Professor Tobias Mayer of Göttingen. The navigator, by measuring the distances and consulting the tables, could, for example, work out the precise time at Greenwich. By comparing this with local time, taken from the movement of the sun (the sextant is fitted with several filters to prevent glare when looking directly at the sun), a navigator could work out his longitude within thirty nautical miles. To achieve this standard, it was necessary to have an instrument that could measure lunar distances more accurately than earlier instruments had managed. The sextant was the answer.

From the collection of Raymond J. and Laura Wielgus.

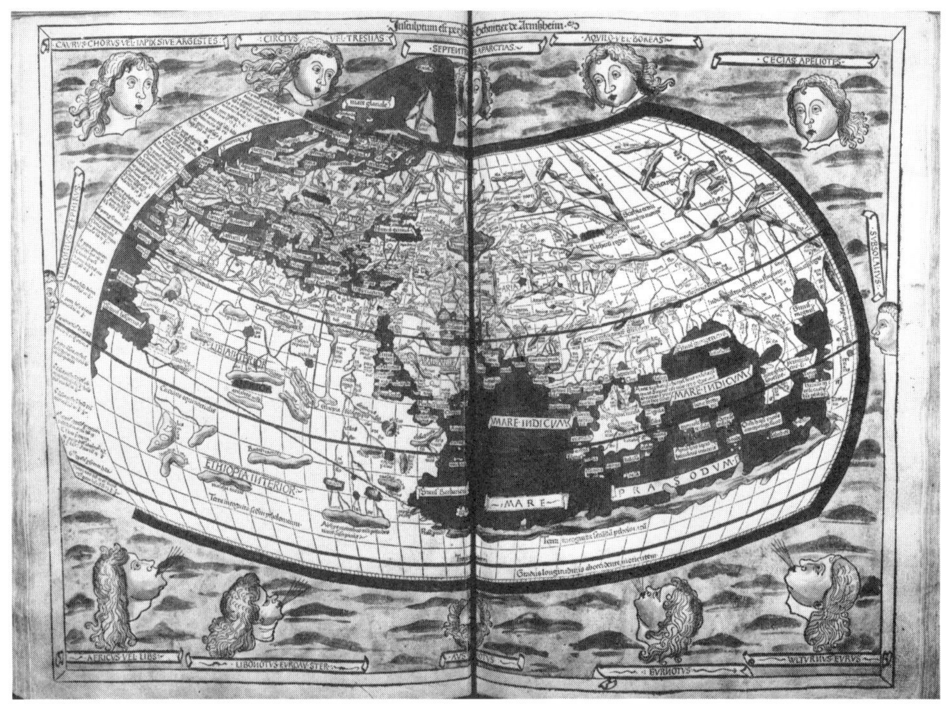

46. Ptolemaeus. *The world map by John Ruysch.*

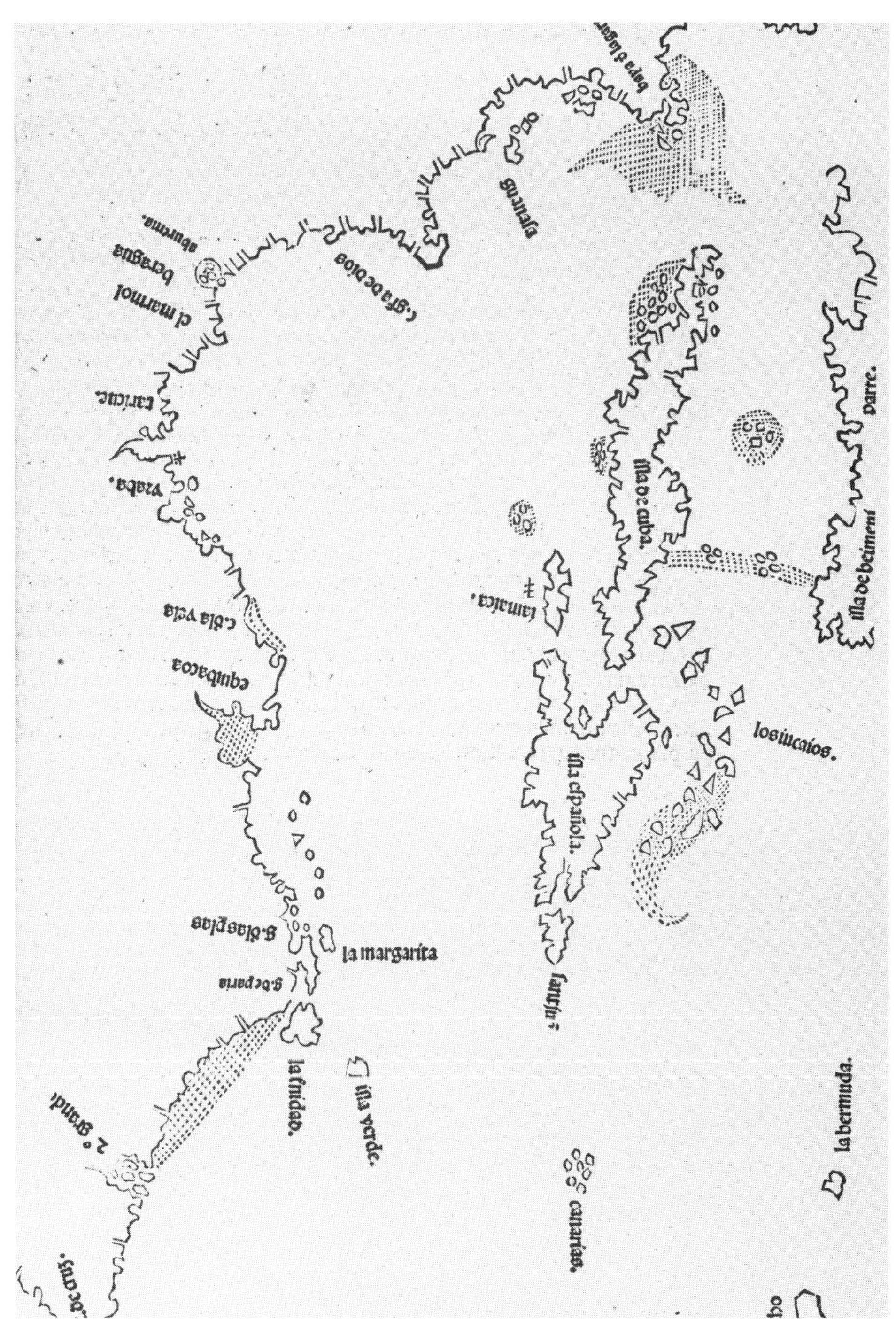

48. Pietro Martire d'Anghiera. *Map of the Americas.*

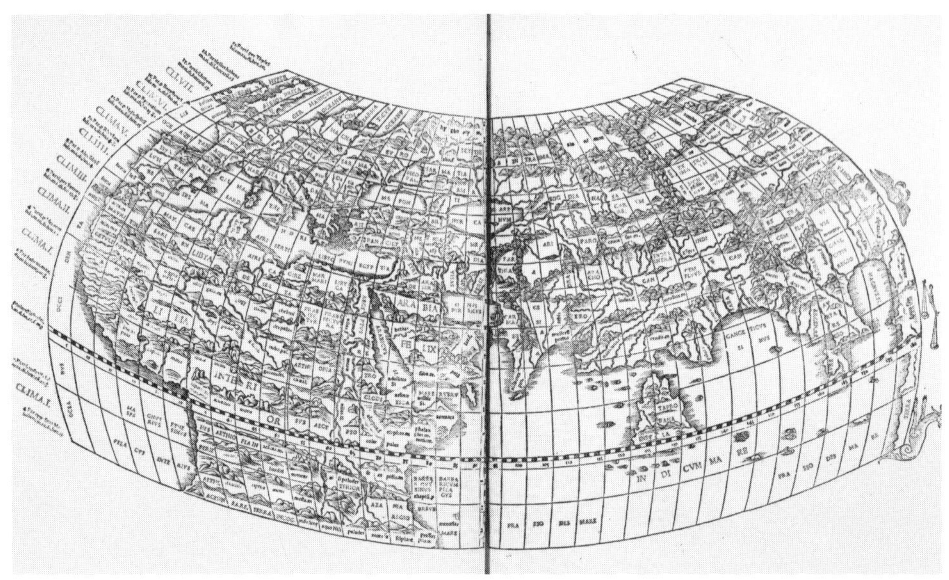

49. Ptolemaeus. *The known world (ecumene) before 1492 comprised the Eurasian and African continents.*

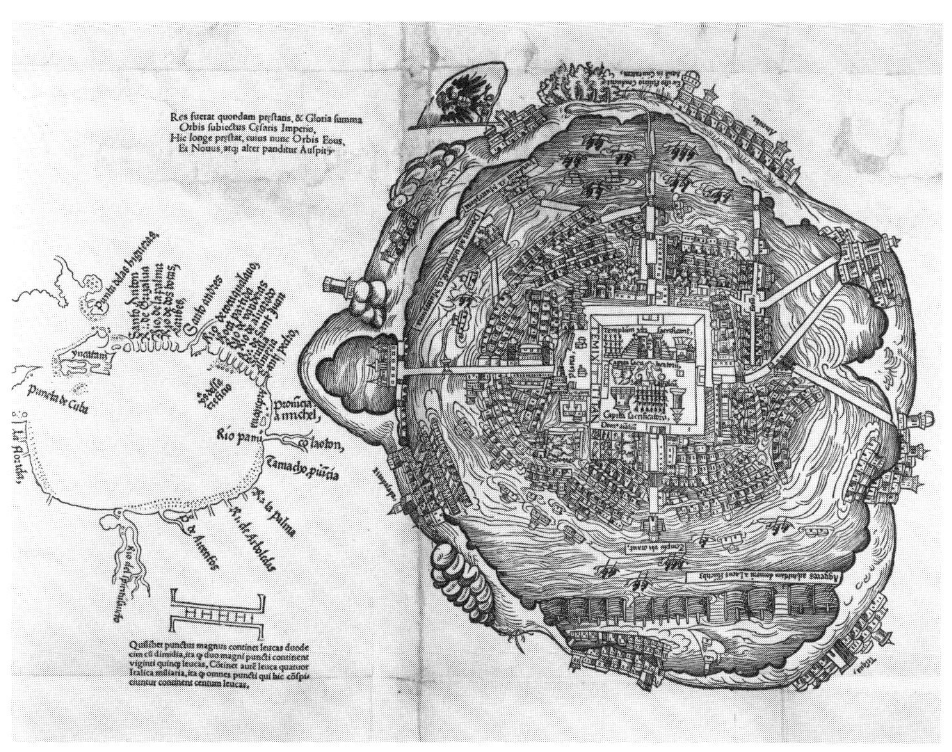

54. Cortés. *Map of Mexico City.*

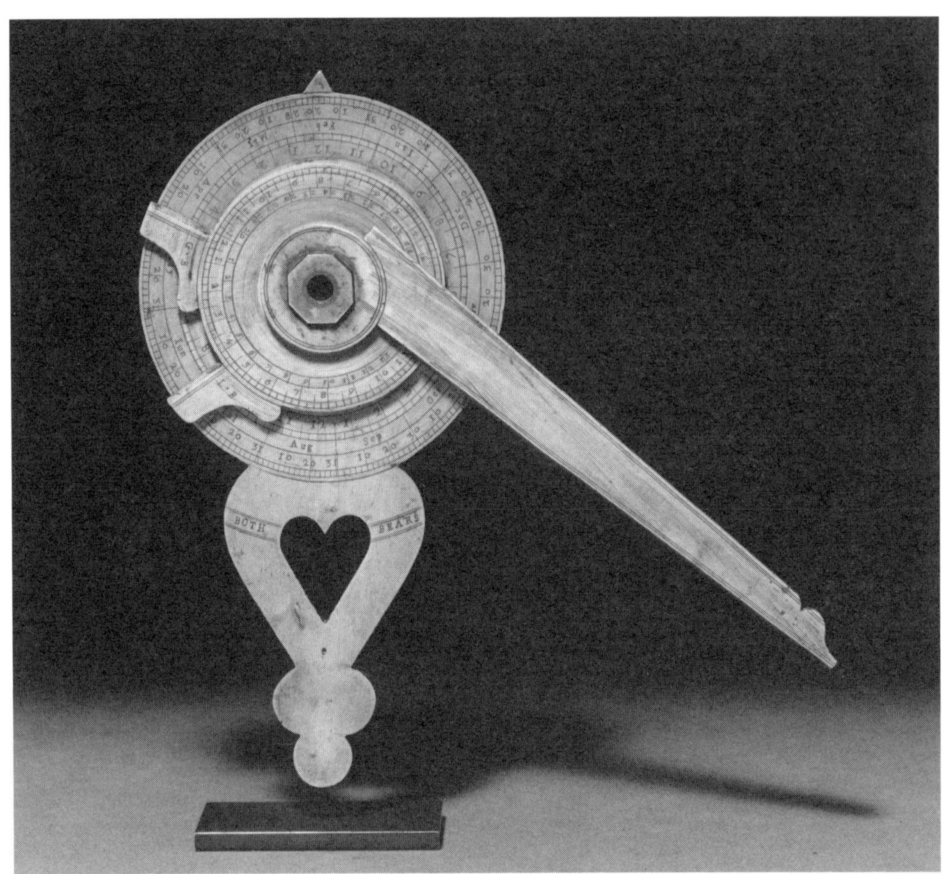

76. Nocturnal.

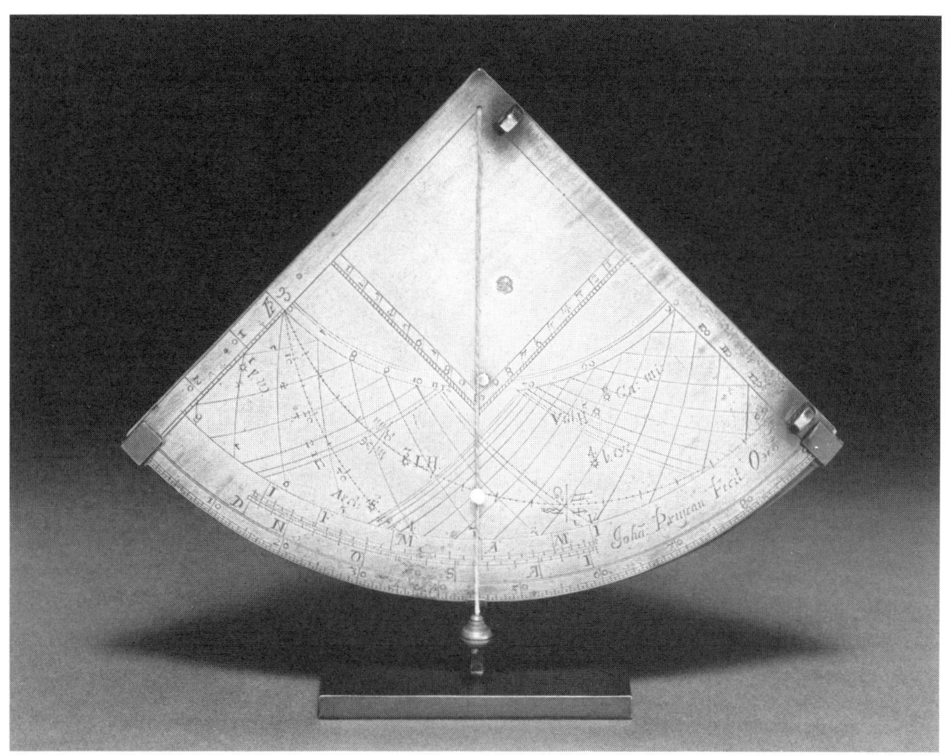

79. Gunter quadrant.

83. Equinoctial ring dial.

Bibliography

Bedini, Silvio, et al. *The Christopher Columbus Encyclopedia.* New York: Simon & Schuster, 1992. 2 vols.

Fernández-Armesto, Felipe. *Before Columbus. Exploration and Colonisation from the Mediterranean to the Atlantic 1229–1492.* London: Macmillan, 1987.

Phillips, William D. and Carla R. Phillips. *The Worlds of Christopher Columbus.* New York: Cambridge University Press, 1992.

Provost, Foster, ed. *Columbus. An Annotated Guide to the Scholarship on His Life and Writings, 1750 to 1988.* Detroit: Omnigraphics for the John Carter Brown Library, 1991.

Randles, W. G. *Portuguese and Spanish Attempts to Measure Longitude in the 16th Century.* Coimbra: 1985.

Russell, Jeffrey Burton. *Inventing the Flat Earth, Columbus and Modern Historians.* New York: Praeger, 1921.

Vigneras, Louis-André. *The Discovery of South America and the Andalusian Voyages.* Chicago: University of Chicago Press, 1976.

These numbers refer to catalogue entries, not catalogue pages.

Lilly Library Publications: Unnumbered Series.

The Dedication of the Lilly Library, October 3, 1960

A. E. Housman

An Exhibit to Commemorate the One Hundred Fiftieth Anniversary of the Beginning of the War of 1812

Exhibition of Original Printings of Some Milestones of Science from Pliny (1469) to Banting (1922)

Grolier: or 'Tis Sixty Years Since

The Upton Sinclair Archives

The Bernardo Mendel Collection

Manuscripts, ancient-modern

The J. K. Lilly Collection of Edgar A. Poe

Three Centuries of American Poetry

Beginning with *Discovery*, exhibition catalogues and other publications from The Lilly Library are numbered consecutively.

I *Discovery*

II *Medicine*

III *One Hundred and Fifty Years ... of Indiana Statehood*

IV *Eighty-Nine Good Novels of the Sea, the Ship, and the Sailor*

V *The First Twenty-Five Years of Printing, 1455–1480*

VI *American Patriotic Songs*

VII *An Exhibit of Seventeenth-Century Editions of Writings by John Milton*

VIII *The Beginnings of Higher Education in Indiana*

IX *The Indiana Wordsworth Collection*

X *Biology. An Exhibition at the Lilly Library in honor of the meetings of The American Institute of Biological Sciences and The American Physiological Society*

Report of the Rare Book Librarian, July 1957–June 1958; July 1958–June 1959; July 1959–June 1961; July 1961–June 1963; July 1, 1963–June 30, 1965; July 1, 1965–June 30, 1967; July 1, 1967–June 30, 1969; July 1980–June 1981; July 1981–June 1982; July 1982–June 1983; July 1983–June 1984; July 1984–June 1985; July 1985–June 1986; July 1986–June 1987; July 1987–June 1988; July 1988–June 1989; July 1989–June 1990; July 1990–June 1991

An Invitation To Join The Friends Of The Lilly Library

The Lilly Library is a central cultural and research institution at Indiana University. Its collections of more than 360,000 books, 6,000,000 manuscripts, and 100,000 pieces of music attract research scholars from around the globe and are available to all who wish to use them.

The Friends of the Lilly Library encourage understanding of the work of the Lilly, bring together enthusiasts of books, libraries, and scholarship, and attract gifts and bequests beyond the normal resources of the Lilly Library.

As patrons of the Lilly's cultural and scholarly activities, members of the Friends of the Lilly Library are recipients of the Lilly exhibition catalogues, and other Lilly Library publications. Friends also receive membership in the Hoosier Book Club of Indiana University Press, which extends a 20 percent discount on all IU Press books. An admirable setting for chamber music concerts, poetry readings, lectures, and special receptions, the Lilly Library opens its doors to many such events sponsored by the Friends.

We hope that you may share an interest in the Lilly Library and express your support through membership in the Friends organization. Following is a list of membership categories:

Regular Membership Categories

Active	$ 25
Sustaining	100
Patron/Corporate	250
Benefactor	500
Life (payable over a five-year period)	1500

Gift Memberships
Contributions may be made as a gift and registered in the recipient's name. A minimum contribution of $25 is required, but may be made in any higher amount as well. The recipient will enjoy all the benefits of Regular Membership.

Memorials
Memorial Fund
A contribution of $25 or more made in the name of an individual will register the recipient in The Friends of the Lilly Library Memorial fund.

Memorial Endowment
A contribution of $5,000 or more will establish a separate memorial endowment in an individual's name. The income from this fund will be used in perpetuity for the purchase of books.

Please make your check payable to Indiana University Foundation/The Friends of the Lilly Library, and send it to the Friends of the Lilly Library, Indiana University, Bloomington, Indiana 47405-3301. Your contribution will qualify as a federal income tax charitable deduction. Indiana residents also qualify for the Indiana Tax Credit.

Indiana University Office of Publications
Editor: Nancy C. McEntire
Designer: Rick Faris

Photography courtesy of Mark Simons, Indiana University Center for Media and
Teaching Resources.